Study Guide with Selected Solutions
for David Moore's

The Basic Practice of Statistics
Fifth Edition

Michael Fligner
The Ohio State University
R. Scott Linder
Ohio Wesleyan University

W.H. Freeman and Company
New York

ISBN-13: 978-1-4292-2783-4
ISBN-10: 1-4292-2783-4

Printed in the United States of America

Second printing

W.H. Freeman and Company
41 Madison Avenue
New York, NY 10010
Houndmills, Basingstoke RG21 6XS, England

www.whfreeman.com

CONTENTS

CHAPTER 1

PICTURING DISTRIBUTIONS WITH GRAPHS

OVERVIEW

Understanding data is one of the basic goals in statistics. To begin, identify the **individuals** or objects described, then the **variables** or characteristics being measured. Once the variables are identified, you need to determine whether they are **categorical** (the variable puts individuals into one of several groups) or **quantitative** (the variable takes meaningful numerical values for which arithmetic operations make sense). The guided solution for Exercise 1.1 provides more details on deciding whether a variable is categorical or quantitative.

After looking over the data and digesting the story behind it, the next step is to describe the data with graphs. Simple graphs give the overall pattern of the data. Which graphs are appropriate depends on whether or not the data are numerical. Categorical data (nonnumerical data) are graphed in **bar charts** or **pie charts.** Quantitative data (numerical data) are graphed in **histograms** or **stemplots.** Quantitative data collected over time use a **time plot** in addition to a histogram or stemplot.

When examining graphs, be on the alert for the following:

• **Outliers** (unusual values) that do not follow the pattern of the rest of the data

• Some sense of a **center** or typical value of the data

• Some sense of how **spread** out or variable the data are

• Some sense of the **shape** of the **overall pattern**

In time plots, be on the lookout for **trends** over time. These features are important whether we draw the graphs ourselves or depend on a computer to draw them for us.

GUIDED SOLUTIONS

Exercise 1.1

KEY CONCEPTS: Individuals and types of variables

(a) When identifying the individual or objects described, you need to include sufficient detail so that it is clear which individuals are contained in the data set.

(b) Recall that the variables are the characteristics of the individuals. Once the variables are identified, you need to determine whether they are categorical (the variable puts individuals into one of several groups) or quantitative (the variable takes meaningful numerical values for which arithmetic operations make sense). Now, list the variables recorded and classify each as categorical or quantitative.

Name of variable   Type of variable

Exercise 1.11

KEY CONCEPTS: Drawing stemplots, splitting stems, and rounding

Hints for drawing a stemplot:

1. It is easiest, although not necessary, to first order the data. If the data have been ordered, the leaves on the stems will be in increasing order. Ordered annual health care spending values follow.

Health Care Spending

419	567	578	682	745	669	838	754	777	1067	1074
1156	1302	1269	1669	1791	1911	1893	1853	1997	2108	2244
2306	2266	2389	2496	2704	2828	2762	2874	2902	2989	3001
2987	3110	3809	3776	5711						

2. Decide how the stems will be shown. Commonly, a stem is all digits except the rightmost. The leaf is then the rightmost digit. Since a stemplot of these data would have many stems and no leaves or just one leaf on most stems, we first **round** the data to the nearest $100. The rounded data follow.

Health Care Spending

400	600	600	700	700	700	800	800	800	1100	1100
1200	1300	1300	1700	1800	1900	1900	1900	2000	2100	2200
2300	2300	2400	2500	2700	2800	2800	2900	2900	3000	3000
3000	3100	3800	3800	5700						

3. Write the stems in increasing order vertically. Write each stem only once, unless you are splitting the stems. In this case, using the stems as the first digits (1000s) would result in the leaves all falling on just a few stems. Because of this, it is best to split the stems. Rounding and splitting are matters of judgment, similar to choosing the classes in a histogram. Now, draw a vertical line next to the stems. Write each leaf (100s column) next to its stem in the plot above. We have included the smallest and largest observations in the partial stemplot below to help you get started.

```
0 4
0
1
1
2
2
3
3
4
4
5
5 7
```

To finish up the exercise, think about the important features that describe a distribution. Does the distribution of the health care spending have a single peak? Does the distribution appear to be symmetric, or is it skewed to the right (tail with larger values is longer) or to the left? What are the center and spread of the distribution and which country is the high outlier?

Exercise 1.25

KEY CONCEPTS: Drawing bar charts

What is the total of the percents in the table? Use the total to compute the percent of vehicles that are some other color.

Complete the following bar chart. The first bar has been drawn for you.

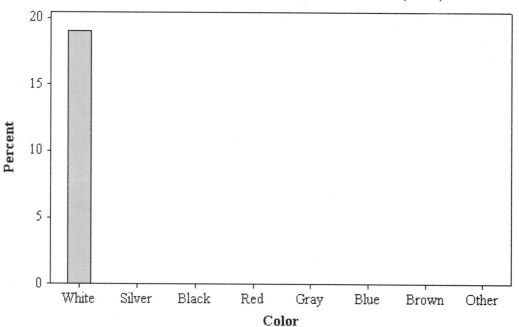

Would a pie chart be appropriate if you added an "Other" category? What about without an "Other" category?

Exercise 1.33

KEY CONCEPTS: Interpreting a histogram

For each of the questions, match the histogram that goes with the variable measured. The first two variables take on only two values and must correspond to figures (b) and (c). Which corresponds to gender and which to handedness? Why? For the last two variables, one has a symmetric distribution and the other is skewed. Which do you think corresponds to height and which to minutes studied?

1) Are you female or male?

2) Are you right-handed or left-handed?

3) What is your height in inches?

4) How many minutes do you study on a typical weeknight?

Exercise 1.45

KEY CONCEPTS: Drawing a histogram, interpreting a histogram, time plots

(a) Following are the ordered values of the number of alligator bites in Florida over the 36-year period from 1972 through 2007. These ordered values will be helpful when counting the number of years in each class interval for the histogram.

```
Alligator Bites
    2     2     3     4     4     5     5     7     7     7     8     9     9     9    10
   12    13    13    13    13    14    15    15    16    17    17    18    18    18    19
   20    20    22    23    23    25
```

When drawing a histogram:

1. Divide the range of values of the data into classes or intervals of equal length.

2. Count the number of data values that fall into each interval.

For this exercise, we are going to use the classes "$2 \leq$ bites < 6," "$6 \leq$ bites < 10," and so on. Complete the table of the number of years in each bite class. If you use software to draw the histogram, the class intervals used may be different. The count for the first class is done for you.

Number of People Bitten by Alligators

Class	Count
2 to <6	7
6 to <10	
10 to <14	
14 to <18	
18 to <22	
22 to <26	

With this small number of data values it is both easy and instructive to draw the histogram by hand. To draw the histogram:

1. Mark the intervals on the horizontal axis and label the axis. Include the units.

2. Mark the scale for the counts or percents on the vertical axis. Label the axis.

3. Draw bars, centered over each interval, up to the height equal to the count or percent. There should be no space between the bars (unless the count for a class is zero, which creates a space between bars).

The vertical axis in a histogram can be either the count or the percent of the data in each interval. Changing the units from counts to percents will not affect the shape of the histogram. In the histogram that follows, the first bar has been drawn for you. Complete the histogram using the information in the table.

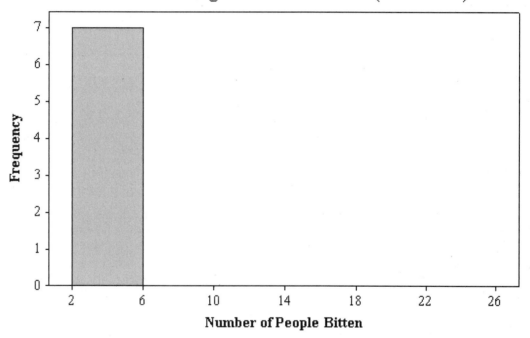

Distribution of Alligator Bites in Florida (1972 - 2007)

What is the midpoint of the yearly counts of people bitten?

(b) Complete the time plot on the graph that follows. The number of alligator bites per year from 1972 to 1985 are drawn on the time plot for you. There are several things to notice when you have completed the graph. There is great variation from year to year and the increasing trend should be apparent. You computed the midpoint of the distribution in (a). How many of the 22 years from 1986 to 2007 had more people bitten than your midpoint from (a)?

Time Plot of Alligator Bites in Florida (1972 - 2007)

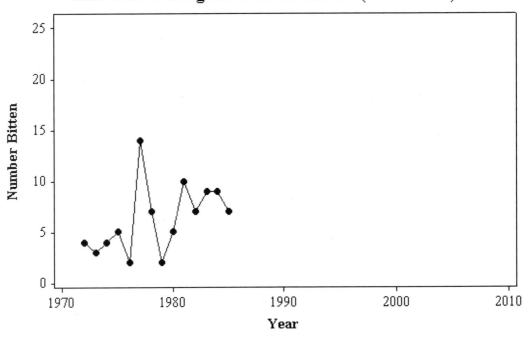

Year

COMPLETE SOLUTIONS

Exercise 1.1

(a) The individuals in this exercise are the different makes and models of cars.

(b) The variables are vehicle type (categorical), transmission type (categorical), number of cylinders (quantitative), city MPG (quantitative), and highway MPG (quantitative). Although the variable "cylinders" is quantitative, this variable divides the cars into only a few categories. To summarize this variable for a group of cars, we could draw a bar graph or a pie chart to give the percentage of 4-, 6-, or 8-cylinder cars in the group, even though these displays are typically used for categorical variables.

Exercise 1.11

Following is the completed stemplot. The distribution is slightly right-skewed with one clear high outlier corresponding to the United States at $5700. Without the United States, the distribution is fairly symmetric. The rounded center of the distribution is around $1900 with about half the countries having annual spending per person less than this value and the other half exceeding $1900. The spread of the distribution goes from $419 to $5711, or to $3809 with the United States removed.

```
0  4
0  66777888
1  11233
1  78999
2  012334
2  578899
3  0001
3  88
4
4
5
5  7
```

Exercise 1.25

The percents in the table sum to $19 + 18 + 16 + 13 + 12 + 12 + 5 = 95$. Thus, the remaining 5% of the vehicles are some other color.

The completed bar chart and pie chart produced by MINITAB follow. If the "Other" category is included, it would be correct to make a pie chart as the categories would now make up the whole range of colors. Without the "Other" category, a pie chart would not be appropriate. A bar chart is appropriate with or without the "Other" category.

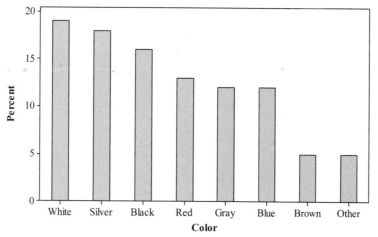

Colors for Vehicles Sold in North America (2007)

Exercise 1.33

1) Are you female or male? This must correspond to either figure (b) or (c). We would expect males and females to be more balanced (closer to 50% in each category) than right-handed or left-handed. So figure (c) should correspond to gender.

2) Are you right-handed or left-handed? This must correspond to either Figure (b) or (c). We know that right-handed people are more common than left-handed people, and since right-handed corresponds to 0, the only possibility is figure (b).

3) What is your height in inches? We would expect a fairly symmetric distribution for heights, which leads us to figure (d).

4) How many minutes do you study on a typical weeknight? This should correspond to a right-skewed distribution with a few people studying much more than the vast majority of students, which leads us to figure (a).

Exercise 1.45

(a) The distribution of the number of people bitten by alligators and the corresponding histogram follow.

Number of People Bitten by Alligators

Class	Count
2 to <6	7
6 to <10	7
10 to <14	6
14 to <18	6
18 to <22	6
22 to <26	4

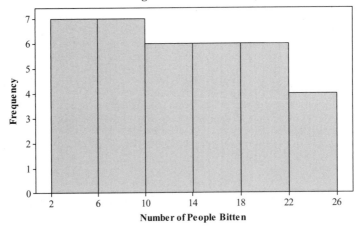

Distribution of Alligator Bites in Florida (1972 - 2007)

The center of the distribution is 13. In about half the years, more than 13 people were bitten by alligators, and in the other half, fewer than 13 were bitten.

(b) The increasing trend over time is apparent in the time plot. In 15 of the 22 years after 1986, more than 13 people were bitten, and in 3 years, exactly 13 people were bitten. Although 13 is the center of the distribution over the 36 years, it does not describe the center of the distribution over the last 22 years. The histogram does not show the increasing trend that is occurring over time.

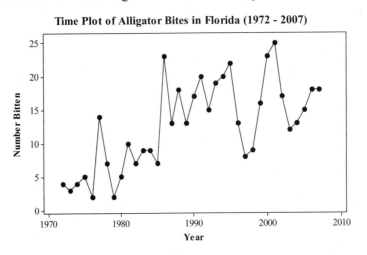

Time Plot of Alligator Bites in Florida (1972 - 2007)

CHAPTER 2

DESCRIBING DISTRIBUTIONS WITH NUMBERS

OVERVIEW

Once you have examined graphs to get an overall sense of the data, it is helpful to look at numerical summaries of features of the data that clarify the notions of center and spread.

Measures of center:
- **mean** (often written as $\overline{x}$)
- **median** (often written as M)

Finding the mean $\overline{x}$

The mean is the common arithmetic average. If there are n observations, $x_1, x_2, \cdots, x_n$, then the mean is

$$\overline{x} = \frac{x_1 + x_2 + \cdots + x_n}{n} = \frac{1}{n}\sum x_i$$

Recall that $\sum$ means "add up all these numbers."

Finding the median M

1. List all the observations from smallest to largest.

2. If the number of observations is odd, then the median is the middle observation. Count from the bottom of the list of ordered values up to the $(n + 1)/2$ largest observation. This observation is the median.

3. If the number of observations is even, then the median is the average of the two center observations.

Measures of spread:
- **quartiles** (often written as Q_1 and Q_3)
- **standard deviation** (s)
- **variance** (s^2)

9

Finding the quartiles Q_1 and Q_3

1.   Locate the median.

2.   The first quartile, Q_1, is the median of the lower half of the list of ordered observations.

3.   The third quartile, Q_3, is the median of the upper half of the list of ordered values.

Finding the variance s^2 and the standard deviation s

1.   Take the average of the squared deviations of each observation from the mean. In symbols, if we have n observations, $x_1, x_2, \cdots, x_n$, with mean $\bar{x}$

$$s^2 = \frac{(x_1 - \bar{x})^2 + (x_2 - \bar{x})^2 + \cdots + (x_n - \bar{x})^2}{n-1} = \frac{1}{n-1}\sum (x_i - \bar{x})^2$$

(Remember, $\sum$ means "add up.")

If you are doing this calculation by hand, it is best to take it one step at a time. First, calculate the deviations, then, square them, next, sum them up, and finally, divide the result by $n - 1$. Exercise 2.10 follows the step-by-step calculations.

2.   The standard deviation is the square root of the variance: $s = \sqrt{s^2}$. Some things to remember about the standard deviation:

(a) s measures the spread around the mean.

(b) s should be used only with the mean, not with the median.

(c) If $s = 0$, then all the observations must be equal.

(d) The larger s is, the more spread out the data are.

(e) s can be strongly influenced by outliers. It is best to use s and the mean only if the distribution is symmetric or nearly symmetric.

For measures of spread, the quartiles are appropriate when the median is used as a measure of center. In fact, the **five-number summary,** reporting the largest and smallest values of the data, the quartiles, and the median, provides a compact description of the data. The five-number summary can be represented graphically by a **boxplot.** If you use the mean as a measure of center, then the standard deviation and variance are the appropriate measures of spread. Watch out because means and variances can be strongly affected by outliers and are harder to interpret for skewed data. The mean and standard deviation are not resistant measures. The median and quartiles are more appropriate when outliers are present or when the data are skewed. The median and quartiles are **resistant measures.**

GUIDED SOLUTIONS

Exercise 2.10

KEY CONCEPTS: Computing the mean, variance, and standard deviation

In practice, you will be using software or your calculator to obtain the mean and standard deviation from keyed-in data. This problem illustrates the step-by-step calculations to help you understand how the computations work. In general, be careful not to round off the numbers until the last step, as too-early rounding can introduce fairly large errors when computing s. However, in this example the numbers work out simply, so there should not be any rounding errors.

(a) Complete the step-by-step computation of the mean:

$$\overline{x} = \frac{1}{n}\sum x_i = \frac{x_1 + x_2 + \cdots + x_n}{n} =$$

(b) To do the step-by-step computation of the standard deviation, first complete the following table to obtain the sum of the squared deviations.

Observations	Deviations	Squared deviations
x_i	$x_i - \overline{x}$	$(x_i - \overline{x})^2$
3175	1036.5	1074331.25
Sum =		Sum =

From the table, we have $\sum(x_i - \overline{x})^2 =$

The variance is then computed as

$$s^2 = \frac{1}{n-1}\sum(x_i - \overline{x})^2 =$$

and the standard deviation as $s = \sqrt{s^2} =$

(c) Enter the data into your calculator and use the appropriate buttons to obtain $\overline{x}$ and s. Since no round-off errors should have been introduced in your hand calculations, the results should agree.

Exercise 2.11

KEY CONCEPTS: Stemplots, mean and standard deviation, shape of distribution

Use statistical software or a calculator to find the mean and standard deviation of the two data sets and write in your answers here.

Data A $\bar{x} =$ $s =$
Data B $\bar{x} =$ $s =$

If you do the calculations correctly, you will find that these are two data sets of 11 observations with the same means and standard deviations. The mean gives an estimate of center, and the standard deviation gives an estimate of spread. Neither of these measures is resistant to outliers, and they do not give an indication of the shape of the distribution. Following is the stemplot of Data A, where the data have been rounded to the nearest 10th. The stems are 1s and the leaves are 10ths. Complete the stemplot of Data B and comment on the shapes of the two distributions.

Data A   Data B

```
3 | 1                                                5 |
4 | 7                                                6 |
5 |                                                  7 |
6 | 1                                                8 |
7 | 3                                                9 |
8 | 1178                                            10 |
9 | 113                                             11 |
                                                    12 |
```

Exercise 2.29

KEY CONCEPTS: Histograms, stemplots and boxplots, mean and standard deviation

If you are unclear on the details of computing the five-number summary, see part (b) of Exercise 2.35, which goes through the step-by-step details. In this exercise, we reproduce the five-number summaries for each variety and ask you to draw the boxplots for the *bihai* and red varieties by hand in the figure on the next page. The boxplot for the yellow variety is drawn for you.

Five-number summaries:

	Bihai	Red	Yellow
Minimum	46.34	37.40	34.57
Q_1	46.71	38.07	35.45
M	47.12	39.16	36.11
Q_3	48.245	41.69	36.82
Maximum	50.26	43.09	38.13

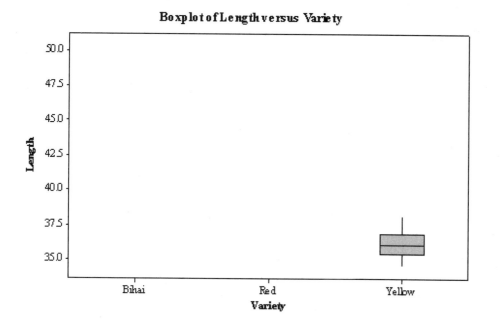

Do the boxplots fail to reveal any important information that is visible in the stemplots?   What information is easily seen in both the stemplots and the boxplots?

Exercise 2.35

KEY CONCEPTS: Describing distributions, measures of center, five-number summary

(a) Histograms, stemplots, and boxplots are three methods for graphing a distribution. First, complete the stemplot below. We have used split stems and the stem units are 100s. The first two stems are completed for you. The first stem contains all days between 0 and 49, with the two leaves corresponding to the observations 43 and 45. The next stem consists of all days between 50 and 99. Note that the 53, 56, 56, 57, and 58 have all been entered as a 5 in the leaves. Complete the stemplot and describe the main features.

```
0  44
0  555556677788888888889999999
1
1
2
2
3
3
4
4
5
5
```

Using the five-number summary found in part (b), complete the boxplot that follows on the next page. Describe the main features of the distribution that are apparent from the boxplot.

Boxplot of Days

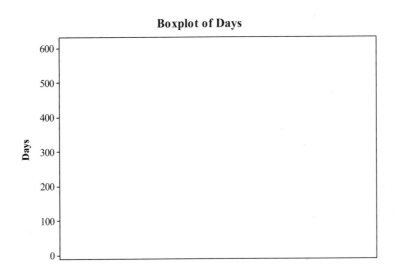

(b) The five-number summary is usually better than the mean and the standard deviation for describing a skewed distribution or a distribution with strong outliers. Because the distribution of days is skewed to the right and may also contain high outliers, we choose the five-number summary that consists of the median, the minimum, the maximum, and the first and third quartiles. To compute these quantities, it is simplest to first order the data, which has already been done. The ordered days are reproduced below. There are 72 observations. What is the location of the median? Compute its value.

43	45	53	56	56	57	58	66	67	73	74	79	80
80	81	81	81	82	83	83	84	88	89	91	91	92
92	97	99	99	100	100	101	102	102	102	103	104	107
108	109	113	114	118	121	123	126	128	137	138	139	144
145	147	156	162	174	178	179	184	191	198	211	214	243
249	329	380	403	511	522	598						

Remember that the first quartile is the median of the 36 observations below the median and that the third quartile is the median of the observations above the median. Determine the quartiles and fill in the following table.

Minimum
Q_1
M
Q_3
Maximum

Explain how you can see from these numbers that the distribution is right-skewed.

The stemplot and boxplot suggests a few large outliers. To determine whether the $1.5 \times IQR$ criterion flags any observations as high outliers, first compute the IQR. It is $Q_3 - Q_1$.

$IQR =$ $1.5 \times IQR =$

Then, add $1.5 \times IQR$ to the third quartile and see if the damage from the largest state exceeds this value.

$$Q_3 + (1.5 \ \times \ IQR) =$$

You should find that there are several suspected outliers in the data according to this criterion.

Exercise 2.39

KEY CONCEPTS: Standard deviation

There are two points to remember in getting to the answer. The first is that numbers "further apart" from each other tend to have higher variability than numbers closer together. The other is that repeats are allowed. There are several choices for the answer to (a) but only one for (b).

COMPLETE SOLUTIONS

Exercise 2.10

(a) The step-by-step computation of the mean is

$$\bar{x} = \frac{1}{n}\sum x_i = \frac{x_1 + x_2 + \cdots + x_n}{n} = \frac{3175 + 2526 + 1763 + 1090}{4} = \frac{8554}{4} = 2138.5 \text{ CFUs per cubic meter of}$$
air.

(b) To organize the calculations in the step-by-step computation of the standard deviation, we first complete this table to obtain the sum of the squared deviations.
S – b

Observations	Deviations	Squared deviations
x_i	$x_i - \bar{x}$	$(x_i - \bar{x})^2$
3175	1036.5	1074331.25
2526	387.5	150156.25
1763	−375.5	141000.25
1090	−1048.5	1099352.25
	Sum = 0	Sum = 2464840.00

From the table, we have $\sum (x_i - \bar{x})^2 = 2464840$. The variance is computed as

$$s^2 = \frac{1}{n-1}\sum (x_i - \bar{x})^2 = \frac{2464840}{4-1} = 821613.33$$

and the standard deviation as $s = \sqrt{s^2} = \sqrt{821613.33} = 906.43$ CFUs per cubic meter of air.

(c) The calculator gives the same results: $\bar{x} = 2138.5$ and $s = 906.43$. Your calculator may vary from this answer depending on the number of significant digits printed. MINITAB prints $\bar{x} = 2139$ and $s = 906$ as its output when printing descriptive statistics for this variable.

Exercise 2.11

Data A	$\bar{x} = 7.501$	$s = 2.031$
Data B	$\bar{x} = 7.501$	$s = 2.031$

From the following stemplots, we see two distributions with quite different shapes but with the same means and standard deviations. Data B seem less spread out than Data A despite the fact that the standard deviations are the same. The reason for this is that the standard deviation is not a resistant measure of spread and its value has been increased by the high outlier in Data B. Data A are left-skewed.

Data A

```
3 | 1
4 | 7
5 |
6 | 1
7 | 3
8 | 1178
9 | 113
```

Data B

```
 5 | 368
 6 | 69
 7 | 079
 8 | 58
 9 |
10 |
11 |
12 | 5
```

Exercise 2.29

The comparative boxplot follows.

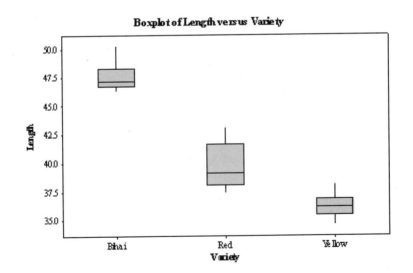

The overall conclusions from the stemplots and boxplots are the same. The *bihai* variety is clearly the longest. All the *bihai* flower lengths are longer than any lengths of the other two varieties. The yellow variety is generally not as long as the red variety. The variability of the yellow and *bihai* varieties seem similar, and the red variety has much more variable lengths than either the yellow or *bihai*. An interesting feature that isn't apparent in the boxplot is the two possible outliers in the *bihai* group that appear in the stemplot. According to the $1.5 \times IQR$ rule, there are *no* suspected outliers.

Exercise 2.35

(a) The completed stemplot and boxplot follow. The right-skew is apparent from both plots with several possible high outliers.

```
0 | 44
0 | 5555566777888888888889999999
1 | 000000000001112222333444
1 | 56777899
2 | 1144
2 |
3 | 2
3 | 8
4 | 0
4 |
5 | 12
5 | 9
```

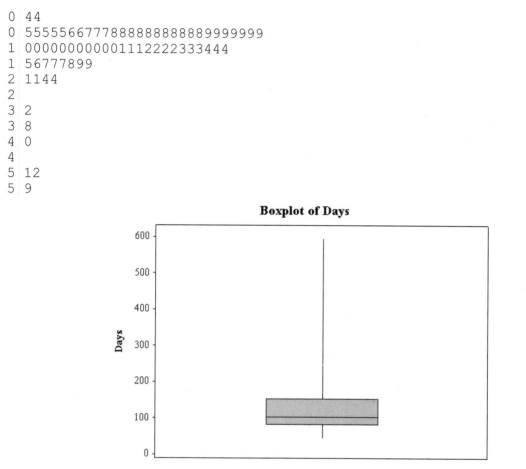

Boxplot of Days

(b) The ordered observations are reproduced in the list that follows. Since there are 72 observations, the median is the average of the two middle observations, which are the 36th and 37th observations. The two middle observations are shaded in gray and underlined. The median is the average of these two observations and is $(102 + 103)/2 = 102.5$.

43	45	53	56	56	57	58	66	67	73	74	79	80
80	81	81	81	*82*	*83*	83	84	88	89	91	91	92
92	97	99	99	100	100	101	102	102	102	103	104	107
108	109	113	114	118	121	123	126	128	137	138	139	144
145	147	*156*	*162*	174	178	179	184	191	198	211	214	243
249	329	380	403	511	522	598						

The first quartile is the median of the 36 observations below the median and is the average of the 18th and 19th largest observation in this group. These two observations are 82 and 83 and are shaded in gray and in italics. Their average is $(82 + 83)/2 = 82.5$, which is the first quartile. Similarly, the third quartile is the average of the 18th and 19th largest observation among the 36 observations above the median. These observations are 156 and 162 and are shaded in gray and in italics. Their average $(156 + 162)/2 = 159$ is the third quartile, and it is also in italics and underlined.

Here is the five-number summary:

Minimum 43
Q_1 82.5
M 102.5
Q_3 159
Maximum 598

In a symmetric distribution, the first and third quartile are equally distant from the median. You can see from the five-number summary that the third quartile is much farther above the median than the first quartile is below it. This is typical of distributions that are skewed to the right. The extremes behave the same way.

The IQR is $159 - 82.5 = 76.5$ and $1.5 \times IQR = 114.75$. If we add 114.75 to the third quartile, we get $159 + 114.75 = 273.75$ days. The observations 329, 380, 403, 511, 522, and 598 are all flagged as suspected high outliers

Exercise 2.39

(a) The standard deviation is always greater than or equal to zero. The only way it can equal zero is if all the numbers in the data set are the same. Since repeats are allowed, just choose four of the same numbers to make the standard deviation equal to zero. Examples are 1, 1, 1, 1 or 2, 2, 2, 2.

(b) To make the standard deviation large, numbers at the extremes should be selected, so you want to put the four numbers at 0 or 10. The correct answer is 0, 0, 10, 10. You might have thought 0, 0, 0, 10 or 0, 10, 10, 10 would be just as good, but a computation of the standard deviation of these choices shows that two at either end is the best choice.

(c) There are many choices for (a) but only one for (b).

CHAPTER 3

THE NORMAL DISTRIBUTIONS

OVERVIEW

This chapter considers the use of **mathematical models** (mathematical formulas) to describe the overall pattern of a distribution. The name given to a mathematical model that summarizes the shape of a histogram is a **density curve.** The density curve is an idealized histogram. The area under a density curve between two x-values represents the proportion of the data that lie between these two numbers. Like a histogram, a density curve can be described by measures of center such as the **median** (a point such that half the area under the density curve is to the left of the point) and the **mean** (the center of gravity or balance point of the density curve). In the last chapter, we called the mean $\bar{x}$. The term refers to the mean of actual observations. The mean of a density curve is referred to as μ. Likewise, the standard deviation of a density curve also has a new notation. It is referred to as σ.

One of the most commonly used density curves in statistics is the **Normal curve.** Normal curves are symmetric and bell-shaped, and as with all density curves, the total area under the Normal curve is 100%. Because the Normal curve is symmetric around the mean, areas (proportions) such as those shown in the figure are equal.

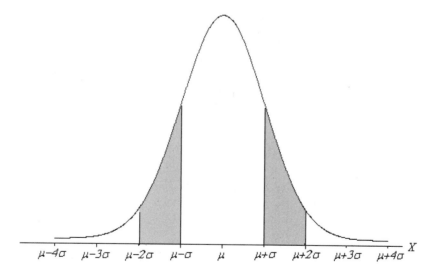

The peak of the Normal curve is above the mean and the standard deviation measures the concentration of the area is around the peak.

Normal curves follow the 68–95–99.7 rule: 68% of the area under a Normal curve lies within one standard deviation of the mean (illustrated in the figure on the next page), 95% within two standard deviations of the mean, and 99.7% within three standard deviations of the mean.

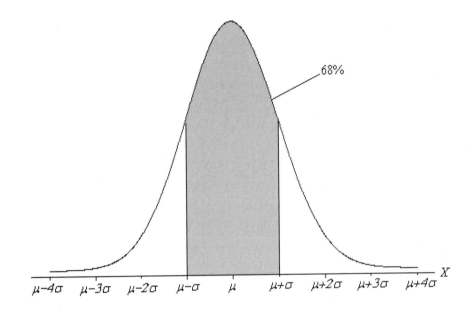

Areas under any Normal curve can be found easily if an *x*-value on the horizontal axis is first standardized by subtracting the mean (μ) from the *x*-value and dividing the result by the standard deviation (σ). This **standardized value** of *x* is called the **z-score.**

$$z = \frac{x - \mu}{\sigma}$$

If data whose distribution can be described by a Normal curve are standardized (all values replaced by their *z*-scores), the distribution of the standardized values is described by the **standard Normal curve.** Areas under standard Normal curves are easily computed by using a **standard Normal table,** such as Table A inside the front cover of your text.

The standard Normal curve is very useful for finding the proportion of observations in an interval when dealing with any Normal distribution. Here are some hints about solving these problems:

    1. State the problem.

    2. Draw a picture of the problem. It will help you know what area you are looking for.

    3. Standardize the observations.

    4. Using Table A inside the front cover of your text, find the area you need.

Use of statistical software: In the Guided Solutions, when looking in the body of Table A for a *z*-value corresponding to a given proportion, we use the value in the table closest to this proportion to solve the problem. We also give the answer that you would obtain if you used statistical software to get the *z*-value corresponding to this proportion.

GUIDED SOLUTIONS

Exercise 3.7

KEY CONCEPTS: 68–95–99.7 rule for Normal density curves

(a) Recall the 68–95–99.7 rule:

68% of the data will be between $\mu - \sigma$ and $\mu + \sigma$,
95% of the data will be between $\mu - 2\sigma$ and $\mu + 2\sigma$, and
99.7% of the data will be between $\mu - 3\sigma$ and $\mu + 3\sigma$.

What are μ and σ in this problem? Use the answer to find the values between which monsoon rains fall in 95% of all years. The following figure should help you visualize the rule.

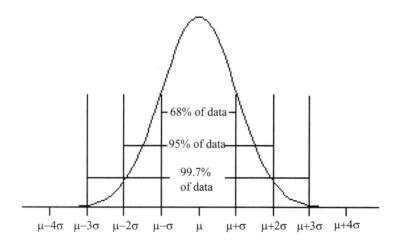

(b) Refer to the figure in (a). Below what value in the figure does 2.5% of the data fall? This value represents the driest 2.5% of all years, and it is one of the values $\mu - 3\sigma$, $\mu - 2\sigma$, or $\mu - \sigma$. Remember that the Normal curve is symmetric about μ.

Now convert this value to a value in millimeters.

Exercise 3.9

KEY CONCEPTS: Standardized scores, z-scores

To compare scores from two Normal distributions, each can be standardized or converted into a z-score. For example, a man's height or a woman's height that corresponds to a z-score greater than 2 places a man in the top 2.5% of men's heights or a woman in the top 2.5% of women's heights. This is because in either case the z-score corresponds to a height that is at least two standard deviations above the mean of its distribution. Using the mean and standard deviation from each distribution, convert a height of 6 feet (72 inches) to a z-score for men and women.

Women z-score =

Men z-score =

Which is more unusual: a 6-foot man or a 6-foot woman? The z-score helps answer this question.

Exercise 3.28

KEY CONCEPTS: Computing areas under a standard Normal density curve

Recall that the proportion of observations from a standard Normal distribution that are less than a given value z is equal to the area under the standard Normal curve to the left of z. Table A gives these areas. The areas are illustrated in the figure.

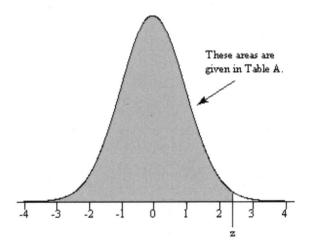

In answering questions concerning the proportion of observations from a standard Normal distribution that satisfy some relation, we find it helpful to first draw a picture of the area under a Normal curve corresponding to the relation. We then try to visualize this area as a combination of areas of the form in the previous figure since such areas can be found in Table A. The entries in Table A are then combined to give the area corresponding to the relation of interest.

This approach is illustrated in the solutions that follow.

(a) To get started, we will work through a complete solution. Following is a picture of the desired area.

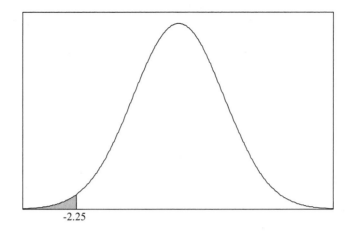

-2.25

This is exactly the type of area that is given in Table A. We simply find the row labeled –2.2 along the left margin of the table, locate the column labeled .05 across the top of the table, and read the entry in the intersection of the row and column: 0.0122. This is the proportion of observations from a standard Normal distribution that satisfies $z < -2.25$.

(b) Shade the desired area in the figure.

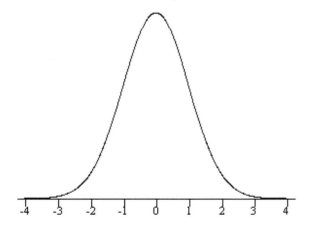

Remembering that the area under the whole curve is 1, how would you modify your answer from part (a)?

Area =

(c) Try solving this part on your own. To begin, draw a picture of a Normal curve and shade the region.

Now, use the same line of reasoning as in part (b) to determine the area of your shaded region. Remember, you want to try to visualize the shaded region as a combination of areas of the form given in Table A.

(d) This part is a bit more complicated than the previous parts, but the same approach will work. Draw a picture and then try to express the desired area as the difference of two regions for which the areas can be found directly in Table A.

Exercise 3.37

KEY CONCEPTS: Finding the value x (the quantile) corresponding to a given area under an arbitrary Normal curve

This is an example of a "backward" Normal calculation. First, we *state the problem and draw a picture*. To make use of Table A on the inside front cover of your text, we need to state the problem in terms of areas to the left of some value. Next, we *use the table*. To do so, we think of having standardized the problem and we then find the value z in the table for the standard Normal distribution that satisfies the stated condition – that has the desired area to the left of it. We next must *unstandardize* this z-value by multiplying by the standard deviation and then adding the mean to the result. The unstandardized value x is the desired result. We illustrate this strategy in the solutions by finding the first quartile, the gas mileage for which the bottom 25% of the vehicles fall below.

State the problem. We are told that the combined city and high gas mileage of 2008 model vehicles are Normally distributed with $\mu = 18.7$ mpg and $\sigma = 4.3$ mpg. We want to find the mileage x that will place a vehicle in the lowest 25% of this distribution. This means that 25% of the vehicles have mileages that are less than x. This is another way of asking you to find the first quartile of the distribution. We first need to find the corresponding value z for the standard Normal. This is illustrated in the following figure.

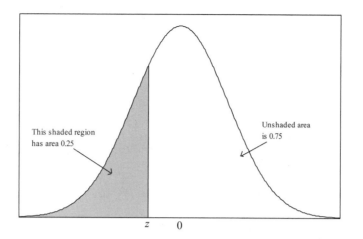

Use the table. The value *z* must have the property that the area to the left of it is 0.25. Areas to the left are the types of areas reported in Table A. Find the entry in the body of Table A that has a value closest to 0.25. This entry is 0.2514. The value of *z* that yields this area is seen from Table A to be –0.67.

Unstandardize. We now unstandardize *z*. The unstandardized value is

$$x = (\text{standard deviation}) \times z + \text{mean} = 4.3z + 18.7 = 4.3 \times (-0.67) + 18.7 = 15.82 \text{ mpg}$$

Thus, a vehicle with gas mileage less than 15.82 mpg is in the lowest 25% of gas mileages.

Statistical software: Using statistical software to get the value of *z* corresponding to 0.25 instead of using the closest value in Table A gives $z = -0.6745$ and $x = 15.80$.

Now, you need to find the mileage so that 75% of the vehicles have a gas mileage below this value. Use the same line of reasoning just illustrated.

State the problem. You may find it helpful to draw the region representing the *z*-value with an area of 75% to the left.

Use the table.

Unstandardize.

Exercise 3.43

KEY CONCEPTS: Computing the area under an arbitrary Normal curve

For these problems, we must first state the problem, then convert the question into one about a standard Normal distribution. This involves standardizing the numerical conditions by subtracting the mean and dividing the result by the standard deviation. We then draw a picture of the desired area corresponding to these standardized conditions and compute the area as we did for the standard Normal distribution, using Table A. This approach is illustrated in the solutions below.

(a) *State the problem*. Call a male SAT score x. The male SAT scores have a $N(533, 116)$ distribution. We want the percent of men with $x > 750$.

  Standardize. We need to standardize the condition $x > 750$. We replace x by z (we use z to represent the standardized version of x) and standardize 750. Since we are told that the mean and standard deviation of SAT scores are 533 and 116, respectively, the standardized value (z-score) of 750 is (rounded to two decimal places).

$$z\text{-score of } 750 = (750 - 533)/116 = 1.87$$

In terms of a standard Normal z, the condition is $z > 1.87$.

  A picture of the desired area follows.

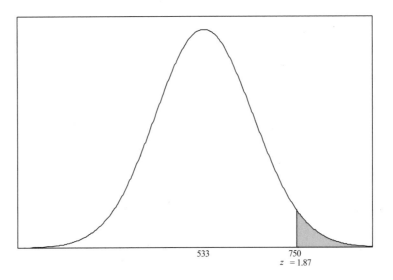

  Use the table. The desired area is not of the form given in Table A. However, we note that the unshaded area to the left of 1.87 is of the form given in Table A and this area is 0.9693. Hence,

shaded area = total area under Normal curve − unshaded area
            = 1 − 0.9693 = 0.0307

Thus, the percent of men whose SAT score x satisfies $x > 750$ is $0.0307 \times 100\% = 3.07\%$, or about 3%.

(b) Try this part on your own. First, *state the problem*.

Next, *standardize*. To do so, compute z-scores to convert the problem to a statement involving standardized values. In terms of z-scores, the condition of interest is,

Standardized condition:

Sketch the standard Normal curve below and shade the desired region on your curve.

Now, *use the table.* Use Table A to compute the desired area. This will be the answer to the question.

COMPLETE SOLUTIONS

Exercise 3.7

(a) In this problem, the mean is $\mu = 852$ millimeters and the standard deviation is $\sigma = 82$ millimeters. From the 68–95–99.7 rule we know that the middle 95% of all monsoon rains should fall between

$$\mu - 2\sigma = 852 - (2 \times 82) = 852 - 164 = 688 \text{ millimeters and}$$

$$\mu + 2\sigma = 852 + (2 \times 82) = 852 + 164 = 1016 \text{ millimeters}$$

(b) Since 95% of all monsoon rains fall between $\mu - 2\sigma = 688$ millimeters and $\mu + 2\sigma = 1016$ millimeters, the remaining 5% of all monsoon rains should be less than 688 millimeters or more than 1016 millimeters. The symmetry of the Normal curve about its mean implies that half of this 5% (in other words, 2.5%) will be below 688 millimeters and the remaining 2.5% above 1016 millimeters. Thus, the monsoon rains in the dryest 2.5% of all years are less than 688 millimeters.

Exercise 3.9

Women $\qquad z\text{-score} = \dfrac{72 - 64}{2.7} = 2.96$

Men $\qquad z\text{-score} = \dfrac{72 - 69.3}{2.8} = 0.96$

Rounding the z-scores of 2.96 and 0.96 to 3 and 1, respectively, and then applying the 68–95–99.7 rule, we see that about 0.15% of women are at least 6 feet tall while about 16% of men are at least 6 feet tall.

Exercise 3.28

(a) A complete solution was provided in the Guided Solutions.

(b) The desired area is indicated in the figure.

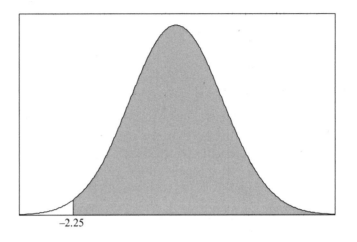

This is not of the form for which Table A can be used directly. However, the unshaded area to the left of −2.25 is of the form needed for Table A. In fact, we found the area of the unshaded portion in part (a). We notice that the shaded area can be visualized as what is left after deleting the unshaded area from the total area under the Normal curve.

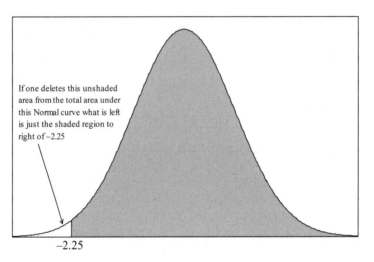

Since the total area under a Normal curve is 1, we have

shaded area   = total area under Normal curve − area of unshaded portion

= 1 − 0.0122 = 0.9878.

Thus, the desired proportion is 0.9878.

(c) The desired area is indicated in the figure.

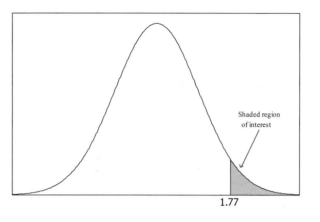

As in part (b) the unshaded area to the left of 1.77 can be found in Table A and is 0.9616. Thus,

shaded area = total area under Normal curve – area of unshaded portion

$$= 1 - 0.9616 = 0.0384$$

This is the desired proportion.

(d) We begin with a picture of the desired area.

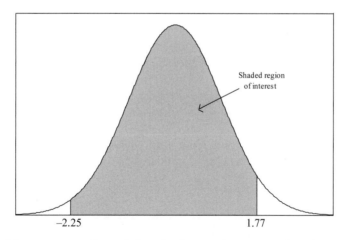

The shaded region is a bit more complicated than in the previous parts, but the same strategy works. We note that the shaded region is obtained by removing the area to the left of –2.25 from all the area to the left of 1.77.

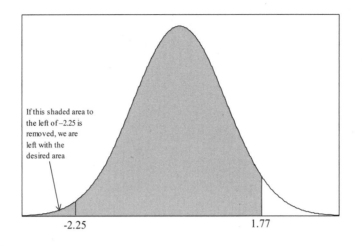

The area to the left of –2.25 is found in Table A to be 0.0122. The area to the left of 1.77 is found in Table A to be 0.9616. Thus,

shaded area = area to left of 1.77 – area to left of –2.25

$$= 0.9616 - 0.0122$$

$$= 0.9494$$

This is the desired proportion.

Exercise 3.37

The complete solution for the lowest 25% (first quartile) of gas mileages is provided in the Guided Solutions. For the lowest 75% (third quartile) we proceed as follows.

State the problem. We are told that the combined city and high gas mileage of 2008 model vehicles are Normally distributed with $\mu = 18.7$ mpg and $\sigma = 4.3$ mpg. We want to find the mileage x that will place a vehicle in the lowest 75% of this distribution. This means that 75% of the vehicles have mileages that are less than x. This is another way of asking you to find the third quartile of the distribution. We first need to find the corresponding value z for the standard Normal. This is illustrated in the figure below.

.

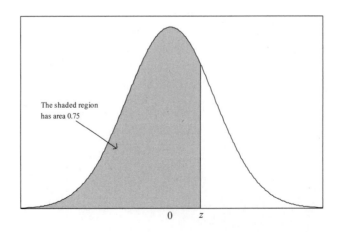

Use the table. The value *z* must have the property that the area to the left of it is 0.75. Areas to the left are the types of areas reported in Table A. Find the entry in the body of Table A that has a value closest to 0.75. This entry is 0.7486. The value of *z* that yields this area is seen from Table A to be 0.67.

Unstandardize. We now unstandardize *z*:

$$x = (\text{standard deviation}) \times z + \text{mean} = 4.3z + 18.7 = 4.3 \times 0.67 + 18.7 = 21.58$$

Thus, 75% of the vehicles have gas mileages that fall below 21.58 mpg. The middle half of the distribution is spanned by the values 15.82 mpg and 21.58 mpg.

Statistical software: Using statistical software to get the value of *z* corresponding to 0.75 instead of using the closest value in Table A gives *z* = 0.6745 and *x* = 21.60 mpg.

Exercise 3.43

(a) A complete solution was provided in the Guided Solutions.

(b) *State the problem.* Call a female SAT score *x*. The female SAT scores have a *N*(499, 110) distribution. We want the percent of women with *x* > 750.

    Standardize. We need to standardize the condition *x* > 750. We replace *x* by *z* (we use *z* to represent the standardized version of *x*) and standardize 750. Since we are told that the mean and standard deviation of SAT scores are 499 and 110, respectively, the standardized value (*z*-score) of 750 is (rounded to two decimal places):

$$z\text{-score of } 750 = (750 - 499)/110 = 2.28$$

In terms of a standard Normal *z*, the condition is *z* > 2.28.

    A picture of the desired area follows.

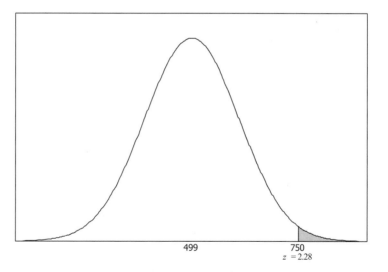

    Use the table. The desired area is not of the form given in Table A. However, we note that the unshaded area to the left of 2.28 is of the form given in Table A and this area is 0.9887. Hence,

shaded area = total area under Normal curve – unshaded area
                = 1 – 0.9887 = 0.0113

Thus, the percent of women whose SAT score exceeds 750 is about 1%, or one-third the percent of the men.

CHAPTER 4

SCATTERPLOTS AND CORRELATION

OVERVIEW

Chapters 1, 2, and 3 of your textbook provide tools for exploring several types of variables one by one. However, in most instances, the data of interest are a collection of variables that may exhibit relationships among themselves. Typically, these relationships are more interesting than the behavior of the variables individually. In this chapter, we consider tools for exploring the relationship between variables. The first tool we consider is the **scatterplot.** Scatterplots display two quantitative variables at a time, such as the weight of a car and its miles per gallon (mpg). Using colors or different symbols, we can add information to the plot about a third variable that is categorical. For example, if in our plot we wanted to distinguish between cars with manual or automatic transmissions, we might use a circle to plot the cars with manual transmissions and a cross to plot the cars with automatic transmissions.

In a scatterplot, one variable is shown on the horizontal axis and the other is shown on the vertical axis. When there is a **response variable** and an **explanatory variable,** the explanatory variable is always placed on the horizontal axis. In cases where there is no explanatory-response variable distinction, either variable can go on the horizontal axis. After drawing the scatterplot by hand or using a computer, the scatterplot should be examined for an **overall pattern** that may tell us about any relationship between the variables and about **deviations** from it. You should be looking for the **direction, form,** and **strength** of the overall pattern. In terms of direction, **positive association** occurs when the variables both take on high values together, while **negative association** occurs if one variable takes on high values when the other takes on low values. In many cases, when an association is present, the variables appear to have a form that can be described as a **linear relationship.** The plotted values seem to form a line. If the line slopes up to the right, the association is positive; if the line slopes down to the right, the association is negative. As always, look for **outliers.** The outlier may be far away from the horizontal variable or the vertical variable or far away from the overall pattern of the relationship.

Categorical variables can be added to a scatterplot by using a different color or symbol for each category. If the response depends on the categorical variable, the overall pattern for points in different categories will differ, and differences will be evident in your plot.

Scatterplots provide a visual tool for looking at the relationship between two variables. Unfortunately, our eyes are not good tools for judging the strength of a relationship. Changes in the scale or the amount of white space in the graph can easily change our judgment of the strength of the relationship. **Correlation** is a numerical measure we use to show the strength of **linear association.**

The correlation can be calculated using the formula

$$r = \frac{1}{n-1}\sum(\frac{x_i - \bar{x}}{s_x})(\frac{y_i - \bar{y}}{s_y})$$

where $\bar{x}$ and $\bar{y}$ are the respective means for the two variables X and Y and s_x and s_y are their respective standard deviations. In practice, you will probably be computing the value of r using computer software or a calculator that finds r from entering the values of the x's and y's. When computing a correlation coefficient, there is no need to distinguish between the explanatory and response variables, even in cases where this distinction exists. The value of r does not change if we switch x and y.

When r is positive, there is a positive linear association between the variables, and when r is negative, there is a negative linear association. The value of r is always between 1 and -1. Values close to 1 or -1 show a strong association, while values near 0 show a weak association. As with means and standard deviations, the value of r is strongly affected by outliers. Their presence can make the correlation much different than it might be with the outliers removed. Finally, remember that the correlation is a measure of straight-line association. There are many other types of association between two variables, but these patterns will not be captured by the correlation coefficient.

GUIDED SOLUTIONS

Exercise 4.1

KEY CONCEPTS: Explanatory and response variables

(a) When examining the relationship between two variables, if you hope to show that one variable can be used to explain variation in the other, remember that the response variable measures the outcome of the study, while the explanatory variable explains changes in the response variable. It is reasonable to assume in this example that the amount of time spent studying for a statistics exam can be used to explain the grade on the exam. Thus, we take the amount of time spent studying as the explanatory variable and the grade as the response variable.

(b) Should you explore the relationship between the variables or view one as explanatory and the other as a response? If you believe you should view one as explanatory and the other as a response,

explanatory variable =   response variable =

(c) Should you explore the relationship between the variables or view one as explanatory and the other as a response? If you believe you should view one as explanatory and the other as a response,

explanatory variable =   response variable =

(d) Should you explore the relationship between the variables or view one as explanatory and the other as a response? If you believe you should view one as explanatory and the other as a response,

explanatory variable =   response variable =

Exercise 4.8

KEY CONCEPTS: Drawing and interpreting a scatterplot

(a) When drawing a scatterplot, one variable (the explanatory variable) is on the horizontal axis and the other (the response) is on the vertical axis. In this data set, we are interested in the "effect" of speed on fuel used. So the speed that the car is driven is the explanatory variable and fuel used is the response variable, shown in the plot in the following figure. Although you will generally draw scatterplots on the computer, drawing a small one like this by hand makes sure that you understand what the points represent. The first two points correspond to speeds of 10 and 20 km/h. Complete the plot by hand.

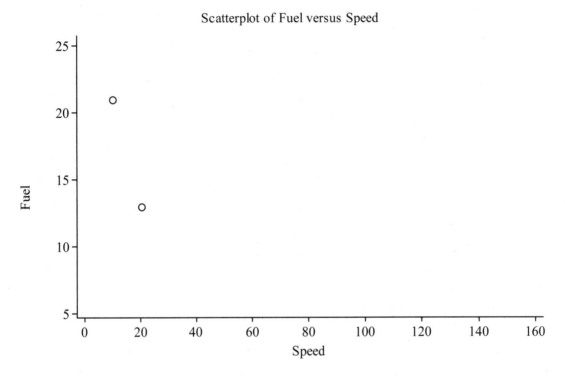

(b) How would you describe the pattern in your plot in words?

Explain why this pattern makes sense. (In your experience, how does fuel consumption vary with speed?)

(c) Positive association occurs when the variables both take on high values together, while negative association occurs if one variable takes on high values when the other takes on low values. Is either of these patterns present here?

(d) How close to a simple curved pattern do the points in the plot lie? If the points lie close to a simple curve, we say that the relationship is strong. If the points are scattered around a curved pattern and do not lie close to it, we say that the relationship is weak.

Exercise 4.9

KEY CONCEPTS: Drawing and interpreting a scatterplot, adding a categorical variable to a scatterplot

(a) When drawing a scatterplot, place one variable (the explanatory variable) on the horizontal axis and the other (the response) on the vertical axis. In this data set, we are interested in the "effect" of Mass on (metabolic) Rate. So Mass is the explanatory variable and Metabolic Rate is the response. To include the categorical variable, Sex, use a different plotting symbol for the two cases (Sex = F and Sex = M). For example, you might use an o for points corresponding to women and an x for points corresponding to men. The points corresponding to the first two women and the first two men are drawn in the plot that follows. Complete the plot, either by hand or by using software.

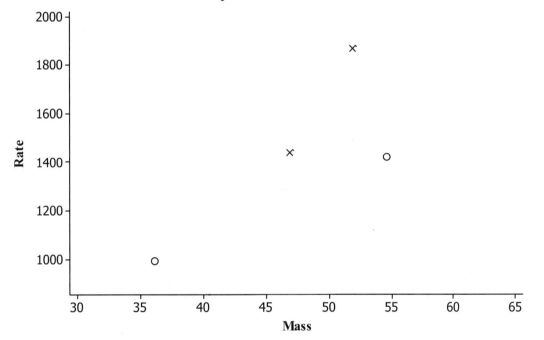

Scatterplot of Rate versus Mass

(b) Describe the direction and strength of the relationship between Mass and Rate for the points corresponding to each Sex. Can these relationships be described with straight lines? How do the patterns for the two runs differ?

Exercise 4.10

KEY CONCEPTS: Scatterplots and computing the correlation coefficient

(a) One might guess that increasing Distance (from initial Ebola outbreak) of a home range will increase the number of Days before the home range is exposed to the virus. Thus, the variable Distance is the explanatory variable, and is plotted on the horizontal axis. Days is the response and is plotted on the vertical axis. Make your scatterplot on the axes provided.

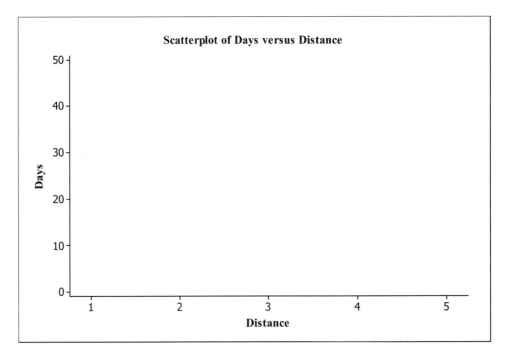

(b) Let x denote Distance and y denote Days. We find the means and standard deviations to be

$$\bar{x} = 3.5 \qquad s_x = 1.3784$$

$$\bar{y} = 31.3333 \qquad s_y = 16.1328$$

Calculations by hand are best done systematically, such as in the following table. The second and fourth columns are the standardized values for x and y. The table entries for the first two x, y values are provided. See if you can complete the remaining entries

x	$\left(\dfrac{x-\bar{x}}{s_x}\right)$	y	$\left(\dfrac{y-\bar{y}}{s_y}\right)$	$\left(\dfrac{x-\bar{x}}{s_x}\right)\left(\dfrac{y-\bar{y}}{s_y}\right)$
1	−1.8137	4	−1.6943	3.0730
3	−0.3627	21	−0.6405	0.2323
4		33		
4		41		
4		43		
5		46		

Now, sum up the values in the last column and divide by $n - 1 = 6{-}1 = 5$ to compute r.

$r =$

(c) Use your calculator to compute r. You may need to consult its operation manual.

Exercise 4.35

KEY CONCEPTS: Changing units of measurement and impact on correlation

(a) You can use software to construct the plot, or you can prepare it by hand on the axes provided.

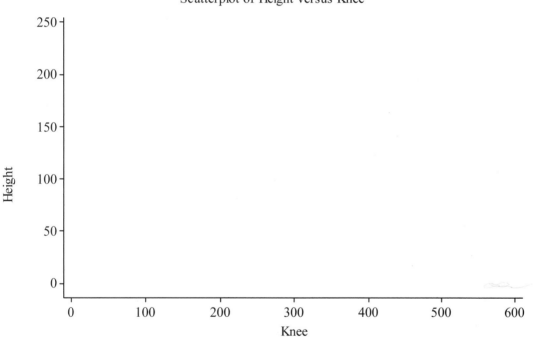

Scatterplot of Height versus Knee

(b) Recall Fact 2, listed in the "Facts about Correlation" section of the chapter. Now, compute the two correlations (software or calculator recommended).

Exercise 4.37

KEY CONCEPTS: Interpreting the correlation coefficient

(a) The key sentence is "A well-diversified portfolio includes assets with low correlations." What does the sentence imply about which of the two investments she should choose?

(b) What can you say about the correlation between two variables when increases in one of the variables are associated with decreases in the other?

Exercise 4.39

KEY CONCEPTS: Drawing and interpreting a scatterplot, computing and interpreting the correlation coefficient

The steps of the process are as follows:

State. What is the practical question, in the context of the real-world setting?

Plan. What specific statistical operations does this problem call for?

Solve. Make the graphs and carry out the calculations needed for the problem.

Conclude. Give your practical conclusion in the setting of the real-world problem.

To apply these steps to this problem, here are some suggestions.

State. The problem asks whether the data support the theory that a smaller percent of birds survive following a successful breeding season. What does the theory suggest about the relationship between breeding pairs and the percent that return the next season?

Plan. What specific statistical operations do you need to do to explore this relationship?

Solve. Draw any plots and compute any quantities specified in the *Plan* step.

Conclude. From the results of the *Solve* step, do the data support the theory that a smaller percent of birds survive following a successful breeding season? Why?

COMPLETE SOLUTIONS

Exercise 4.1

(a) A complete solution is provided in the Guided Solutions.

(b) We would probably simply want to explore the relationship between weight and height.

(c) We would probably view hours per week of extracurricular activities as explaining grade point average. Thus, the response is grade point average and the explanatory variable is hours of extracurricular activities per week.

(d) We would probably simply want to explore the relationship between a student's scores on the SAT math exam and student's SAT verbal exam.

Exercise 4.8

(a) The complete scatterplot follows.

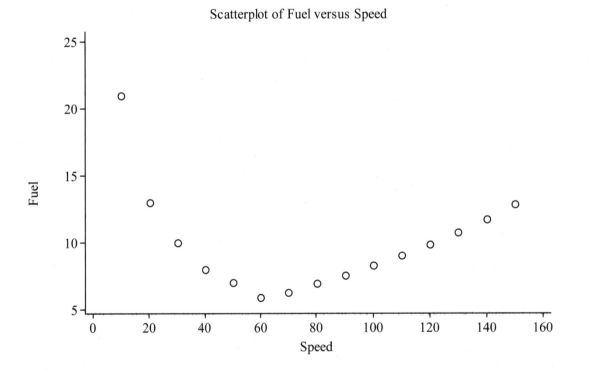

Scatterplot of Fuel versus Speed

(b) The plot shows a curved relationship. Fuel used first decreases as speed increases, and then at about 60 km/h increases as speed increases. This is not surprising. At very slow speeds and at very high speeds, engines are very inefficient and use more fuel, while at moderate speeds, engines are more efficient and use less fuel.

(c)   Variables are positively associated when both take on high values together and both take on low values together. Negative association occurs when high values of one variable are associated with low values of the other. In the scatterplot, both low and high speeds correspond to high values of fuel used, so we cannot say that the variables are positively or negatively associated.

(d)   The points appear to lie close to a simple curve, so we say the relationship is reasonably strong.

Exercise 4.9

(a) The scatterplot for all the data follows.

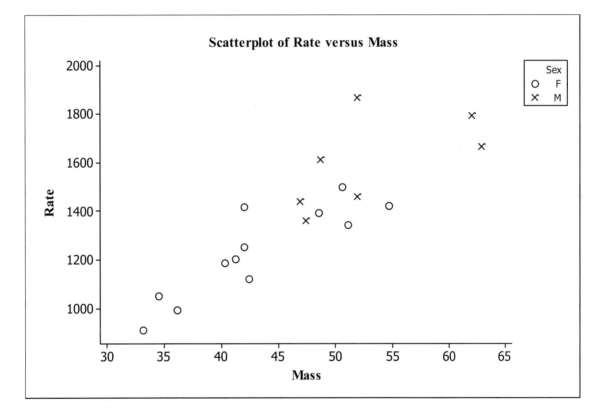

(b)   For both sexes, there is a reasonably strong straight line relationship between metabolic rate and mass. In general, larger people have greater metabolic rates. The rates of growth in metabolic rate with mass is similar for both sexes, but clearly most men are more massive than most women (and have higher metabolic rates).

Exercise 4.10

(a) The scatterplot follows.

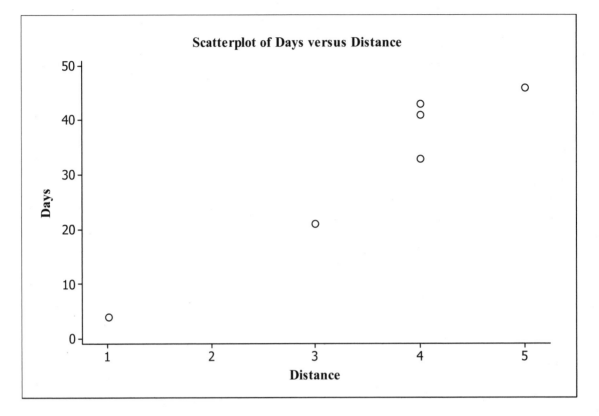

Although not perfect, the points do appear to lie approximately on a straight line. Certainly, there is a strong linear relationship between Days and Distance.

(b) The completed table follows.

x	$\left(\dfrac{x-\overline{x}}{s_x}\right)$	y	$\left(\dfrac{y-\overline{y}}{s_y}\right)$	$\left(\dfrac{x-\overline{x}}{s_x}\right)\left(\dfrac{y-\overline{y}}{s_y}\right)$
1	−1.8137	4	−1.6943	3.0730
3	−0.3627	21	−0.6405	0.2323
4	0.3627	33	0.1033	0.0375
4	0.3627	41	0.5992	0.2173
4	0.3627	43	0.7218	0.2618
5	1.0882	46	0.9074	0.9874

The sum of the values in the last column is 4.8093. Thus, the correlation is
$$r = 4.8093/5 = .9619$$
A correlation this close to 1 suggests that there is a strong positive linear relationship between Days and Distance. This is in agreement with the plot in (a.).

(c) The calculator gives a correlation of 0.9623, which agrees with the result in part (except for rounding).

Exercise 4.35

(a) The scatterplot follows. The points represented by the symbol o are the original data, and the points represented by the symbol x are those of the mad scientist. The patterns for the two sets of data look very different.

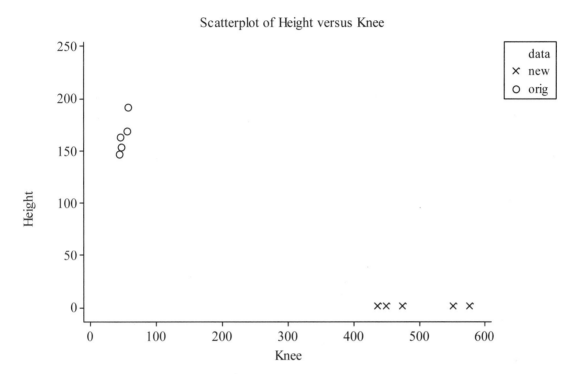

Scatterplot of Height versus Knee

(b) According to Fact 2 in the section on facts about correlation in your textbook, "Because r uses the standardized values of the observations, r does not change when we change the units of measurement of x, y, or both." Thus, we know without doing any calculations, that the correlation is exactly the same for the two sets of measurements. If you calculate the correlation for each, you find that both have a correlation of 0.877.

Exercise 4.37

(a) Rachel should be looking for investments that have a weak correlation with municipal bonds, so she should invest in the small-cap stocks. The correlation between municipal bonds and the small-cap stock is weaker (closer to 0) than the correlation between municipal bonds and the large-cap stocks.

(b) When increases in one variable are associated with decreases in another, the two variables are negatively associated and the correlation is negative. Thus, Rachel should look for assets that are negatively correlated with municipal bonds. Neither large-cap nor small-cap stocks meet this criterion.

Exercise 4.39

State. A successful breeding season is indicated when the number of breeding pairs is high. The percent that return the next season is a measure of survival. According to the theory, if the number of breeding pairs is high (above average), the percent that return the next season will be low (below average). To see whether the data support the theory, we should examine the relationship between the number of breeding pairs and the percent that return the next breeding season.

Plan. We should take the number of breeding pairs as the explanatory variable and the percent that return the next season as the response. We should make a scatterplot of the data to display the relationship between the variables and see if it supports the theory. If the relationship appears to be approximately linear, we should compute the correlation coefficient to quantify the strength of the relationship.

Solve. A scatterplot of the data follows.

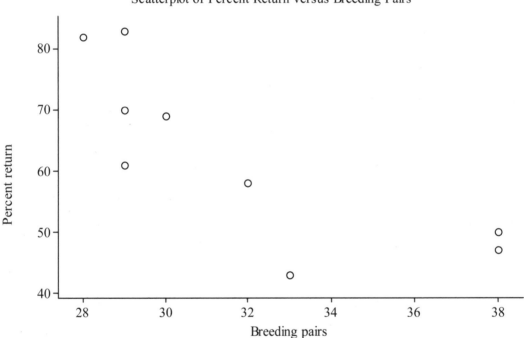

The correlation coefficient is $r = -0.794$.

Conclude. The scatterplot shows a moderately strong negative association. The trend is roughly linear (although you may see a slight curve). The correlation coefficient is -0.794, confirming that the association is negative and that the relationship is moderately strong. The data support (or are consistent with) the theory that a smaller percent of birds survive following a successful breeding season.

CHAPTER 5

REGRESSION

OVERVIEW

If a scatterplot shows a linear relationship that is moderately strong as measured by the correlation, we can draw a line on the scatterplot to summarize the relationship. In the case where there is a response and an explanatory variable, the **least-squares regression** line often provides a good summary of the relationship. A straight line relating y to x has the form

$$y = a + bx$$

where b is the **slope** of the line and a is the **intercept.** The slope tells us the change in y corresponding to a one-unit increase in x. The intercept tells us the value of y when x is 0. This information has no practical meaning unless 0 is a feasible value for x.

The least-squares regression line is the straight line $\hat{y} = a + bx$, which minimizes the sum of the squares of the vertical distances between the line and the observed values y. The formula for the slope of the least-squares line is

$$b = r\frac{s_y}{s_x}$$

and for the intercept is $a = \bar{y} - b\bar{x}$, where $\bar{x}$ and $\bar{y}$ are the means of the x and y variables, s_x and s_y are their respective standard deviations, and r is the value of the correlation coefficient. Typically, the equation of the least-squares regression line is obtained by computer software or a calculator with a regression function. The least-squares regression line can be used to predict the value of y for any value of x. Just substitute the value of x into the equation of the least-squares regression line to get the predicted value for y.

Correlation and regression are clearly related, as can be seen from the equation for the slope b. However, the more important connection is how r^2 measures the strength of the regression. The square of the correlation coefficient, r^2, tells us the fraction of the variation in y that is explained by the regression of y on x. The closer r^2 is to 1, the better the regression describes the connection between x and y.

An examination of the **residuals** shows us how well our regression does in predictions. The difference between an observed value of y and the predicted value obtained by least-squares regression, $\hat{y}$, is called the residual.

$$\text{residual} = y - \hat{y}$$

Plotting the residuals is a good way to check the fit of a least-squares regression line. Features to look for in a **residual plot** are unusually large values of the residuals (outliers), nonlinear patterns, and uneven variation about the horizontal line through zero (corresponding to uneven variation about the regression line). Also look for influential observations. **Influential observations** are individual points whose removal would cause a substantial change in the regression line. Influential observations are often outliers in the horizontal direction.

Correlation and regression must be interpreted with caution.

  • Do not **extrapolate.** To extrapolate means to predict beyond the range of the data.

  • Be aware of possible **lurking variables,** variables that have an important effect on the relationship among the variables in a study but are not included among the variables studied.

  • Are the data averages or from individuals? **Averaged data** usually lead to overestimating the correlations.

  • Remember that *association is not causation!* Just because two variables are correlated doesn't mean that one causes changes in the other. The best evidence that an observed association is due to causation comes from a carefully designed experiment.

GUIDED SOLUTIONS

Exercise 5.4

KEY CONCEPTS: Drawing and interpreting the least-squares regression line, least-squares regression, prediction

(a) Use a calculator or statistical software to compute r. Write down your result.

$$r =$$

(b) The following results were obtained using the Minitab software package.

```
The regression equation is
Rate = 201 + 24.0 Mass


Predictor    Coef    SE Coef    T      P
Constant     201.2   181.7     1.11   0.294
Mass         24.026    4.174   5.76   0.000


S = 95.0808   R-Sq = 76.8%   R-Sq(adj) = 74.5%

Analysis of Variance

Source         DF     SS      MS      F      P
Regression      1   299551  299551  33.13  0.000
Residual Error 10    90404    9040
Total          11   389955
```

The equation of the least-squares regression line is given at the beginning of the output. The intercept and slope can also be found in the table whose first column is labeled "Predictor." The column headed "Coef" gives the intercept (on the row labeled "Constant") and the slope (on the row labeled "Mass") in the same column. The values in this table are given to more decimal places than are in the equation at the beginning of the output.

From the output, we see that the equation of the least-squares regression line is

$$\text{Rate} = 201 + 24.0 \times (\text{Mass})$$

To make a scatterplot of the data, use statistical software or the axes provided. Some software will also draw the least-squares regression line on the plot for you.

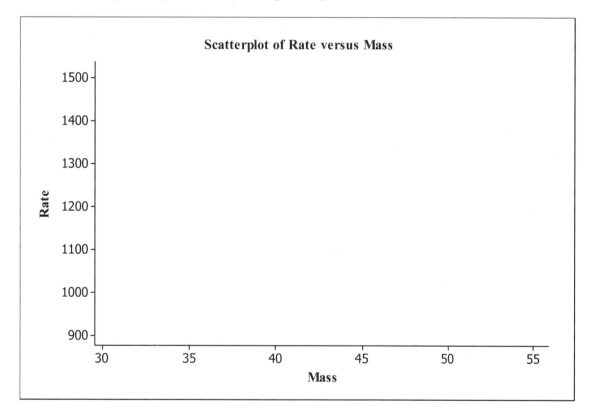

After plotting the data, next plot the regression line on this graph. Perhaps the simplest way to draw a graph of the least-squares regression line is to pick two convenient values for the variable Mass, substitute them into the equation of the least-squares regression line, and for each value, compute the corresponding value of Rate predicted by the equation. This process produces two points on the regression line. Simply plot these points on the graph and connect them with a straight line.

Convenient values for Mass might be 35 and 50. Find the following values.

For Mass = 35, predicted Rate =

For Mass = 50, predicted Rate =

Plot these two sets of values on the graph. Then connect them with a straight line.

(c) The slope of the least-squares regression line is 24.0. Explain clearly in writing what this slope says about the change in predicted metabolic Rate for each 1 kilogram increase in mass.

(d) To predict the metabolic rate of a 45 kilogram woman, substitute 45 for Mass in the equation of the least-squares regression line:

$$\text{Rate} = 201 + 24.0 \times (\text{Mass}) = 201 + 24.0 \times (\quad) = $$

Exercise 5.6

KEY CONCEPTS: Drawing a scatterplot, adding the least-squares regression line to the plot, computing and interpreting r^2

(a) We discussed how to prepare the requested plot in the Guided Solution to Exercise 5.4. Use statistical software or the axes provided to create the plot, then superimpose the regression line.

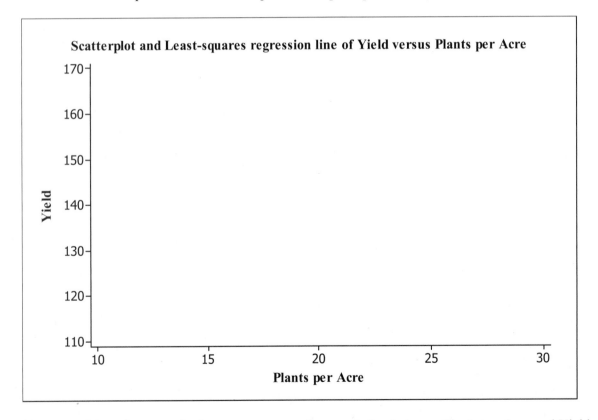

(b.) Use software or your calculator to compute r, the correlation between Plants per Acre and Yield. Use this value to compute r^2.

Exercise 5.8

KEY CONCEPTS: Drawing a scatterplot, adding the least-squares regression line to the plot, prediction, residuals, plotting the residuals

(a) Use statistical software or the axes provided to create the plot. Then superimpose the regression line.

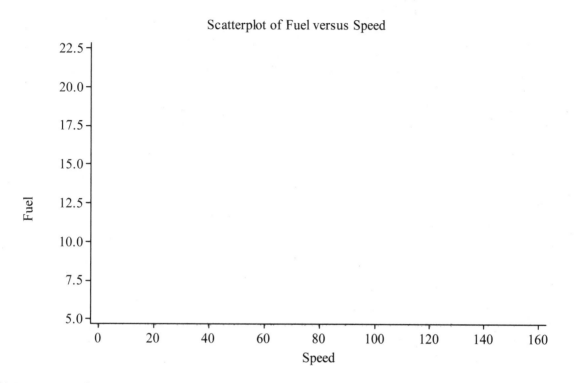

(b) From your plot in (a), does it look as though the fitted regression line fits the data well?

(c) We know that

$$\text{residual} = \text{observed } y - \text{predicted } y$$

For x (speed) = 10, what is the observed value of y (fuel) in the data?

$$y =$$

For $x = 10$, compute the predicted value of y using the least-squares regression line:

$$\hat{y} = 11.058 - 0.01466x =$$

Now compute the residual: $\text{residual} = y - \hat{y} =$

Use a calculator or software to add up the values of the residuals given in the problem.
                                sum =

(d) Use statistical software or the axes provided to create a residual plot. Remember to add a horizontal line to the plot at height 0.

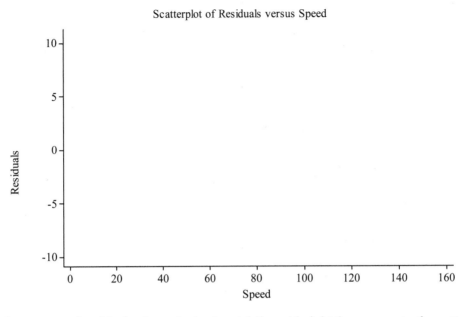

How does the pattern of residuals about the horizontal line at height 0 compare to the pattern of the data points about the least-squares regression line in part (a)?

Exercise 5.10

KEY CONCEPTS: Outliers, influential observations

(a) Use statistical software or the axes provided to create the plot. Note that in the Guided Solution for Exercise 5.4, you made a scatterplot of the original 12 points. You may want to use that plot instead of drawing a new plot. Be sure to add points A and B to your plot. Use different plotting symbols to identify the points.

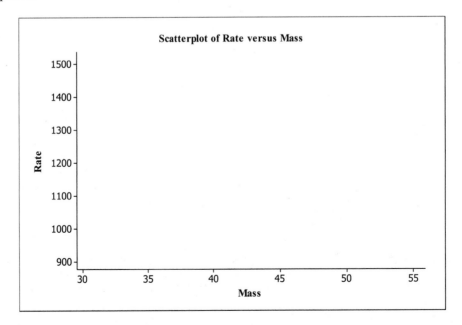

In which direction is each new point an outlier?

Point A:

Point B:

(b) Use a calculator or statistical software to find the least-squares regression line for each of the following.

• Original 12 points

      Least-squares regression line:

• Original 12 points plus Point A

      Least-squares regression line:

• Original 12 points plus Point B

      Least-squares regression line:

Add these three lines to your scatterplot in part (a).
Which new point is more influential?

Why do the points move the line in the way your graph shows? Think about the direction of each outlier.

Exercise 5.12

KEY CONCEPTS: Least-squares regression, prediction, extrapolation

(a) Which variable (Boats or Manatees) is the response variable? Use a software package or your calculator to compute the least-squares regression line for predicting manatees killed from the number (in thousands) of boats registered in Florida.

Write the regression equation here:

(b) Use the equation of the least-squares regression line computed in part (a) to predict the number of manatees killed. Remember that the data you entered into the computer involved number of boats recorded in thousands. Hence, 1,027,000 boats would translate to Boats = 1027.

Is this prediction trustworthy? Our regression line is computed from data where number of boats ranged from 447 to 1024. If Boats = 1027, is this far outside of this range?

(c) Use the equation of the least-squares regression line you computed in part (a) to answer this question.

Is this prediction trustworthy? Refer to the hint provided in (b) above.

Exercise 5.32

KEY CONCEPTS: Calculating the least-squares regression line from summary statistics, r and r^2, prediction

(a) Recall that if the least-squares regression line has the equation

$$\hat{y} = a + bx$$

the formula for the slope of the least-squares line is

$$b = r\frac{s_y}{s_x}$$

and the formula for the intercept is

$$a = \bar{y} - b\bar{x}$$

where $\bar{x}$ and $\bar{y}$ are the means of the x and y variables, s_x and s_y are their respective standard deviations, and r is the value of the correlation coefficient.

What are the x and y variables in this problem?

Use the values for $\bar{x}$, $\bar{y}$, s_x, s_y, and r given in the problem to compute the slope b and intercept a.

$b =$ $a =$

Equation of least-squares regression line: $\hat{y} =$

(b) In part (a) you should have found that the equation of the least-squares regression line is

$$\text{final-exam score} = 30.2 + 0.16 \times \text{(pre-exam total)}$$

Use this equation to predict the final-exam score for a student with a pre-exam total of 300 points.

   predicted final-exam score =

(c) What quantity tells you the fraction of the variation in the values of y (in this case final-exam score) that is explained by the least-squares regression of y (or final-exam score) on x (in this case pre-exam total)? Compute this quantity. Remember to convert the fraction to a percent. What does this calculation tell you about the accuracy of the prediction?

Exercise 5.42

KEY CONCEPTS: Lurking variable

Think about what sort of person is likely to use artificial sweeteners. What is a lurking variable that prevents us from concluding that artificial sweeteners cause weight gain?

Exercise 5.54

KEY CONCEPTS: Scatterplots, least-squares regression, correlation, r^2, the four-step process

The four-step process involves the following steps:

   State. What is the practical question in the context of the real-world setting?

   Plan. What specific statistical operations does this problem call for?

   Solve. Make the graphs and carry out the calculations needed for the problem.

   Conclude. Give your practical conclusion in the setting of the real-world problem.

Here are some suggestions for applying the steps to this problem.

State. What does the problem ask you to do in the context of the real-world setting?

Plan. Is there a graph that would help you uncover the *nature* of the effect of decreasing snow cover on wind stress?

What numerical quantities describe the *strength* of the effect of decreasing snow cover on wind stress?

Solve. Make graphs and calculate any quantities you identified in the *Plan* step.

Conclude. What do you conclude about the nature and strength of the effect of decreasing snow cover on wind stress?

COMPLETE SOLUTIONS

Exercise 5.4

(a) Using software, we find the correlation between Mass and Rate to be

$$r = .876$$

(b) The equation of the least-squares regression line is given in the Guided Solution. A scatterplot with the regression line superimposed follows.

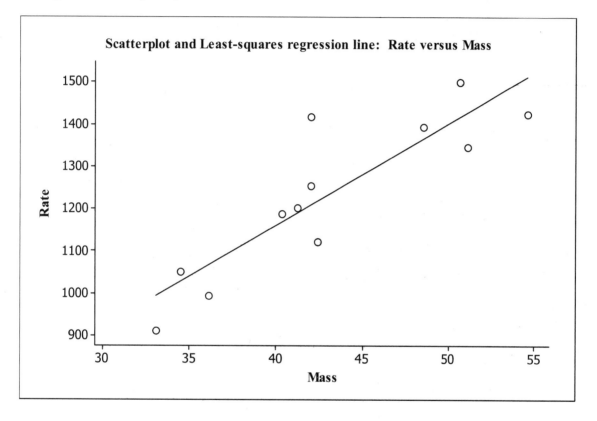

(c) In the Guided Solution, we found that Slope = 24.0. This tells us that an increase of 1 kilogram in mass corresponds to an increase of 24 calories in metabolic rate.

(d) Predicted rate = 201 + 24.0 × (Mass) = 201 + 24.0 × (45) = 1281 calories. We would predict a 45kg woman to have a metabolic rate of 1281 calories per day.

Exercise 5.6

(a) A scatterplot of these data is given, along with the least-squares regression line.

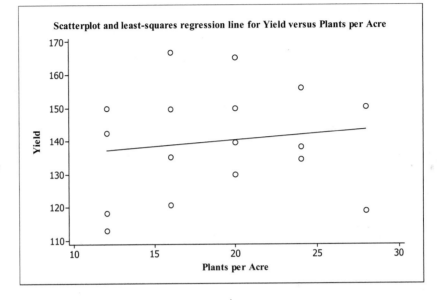

We see that there is not a strong linear relationship between Yield and Number of plants per acre. That is, plants per acre does a poor job of predicting yield.

(b) From software or a calculator, $r = .135$, so $r^2 = (.135)^2 = .018$. This means that less than 2% (1.8%) of the variability in Yield from plot to plot is explained by regression on the number of plants per acre. Number of plants per acre tells us little to nothing about yield, and is therefore not useful for making predictions.

Exercise 5.8

(a) Using statistical software, we obtain the following plot.

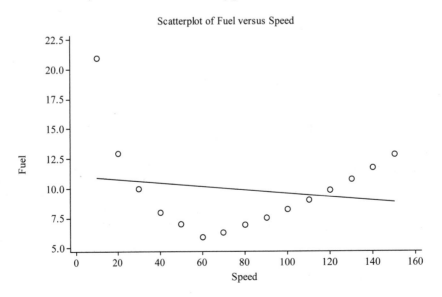

(b) The data points clearly do not follow a straight line pattern. Thus, the least-squares regression line does not fit the data very well. As a consequence, I would not use the least-squares regression line to predict y from x.

(c) For x (speed) = 10, the observed value of y (fuel) in the data is

$$y = 21.00$$

For $x = 10$, compute the predicted value of y using the least-squares regression line:

$$\hat{y} = 11.058 - 0.01466x = 11.058 - 0.01466 \times 10 = 10.91$$

Now compute the residual:

$$\text{residual} = \text{observed } y - \text{predicted } y = 21.00 - 10.91 = 10.09$$

Using a calculator, we find that

$$\text{sum of residuals} = -0.01$$

which differs from 0 only due to roundoff error.

(d) Using software, we obtain the following plot. The pattern of points about the horizontal line at height 0 is almost identical to that of the data points about the least-squares regression line in part (a). The difference is that the pattern in this plot is tilted slightly compared to the pattern in the part (a) plot.

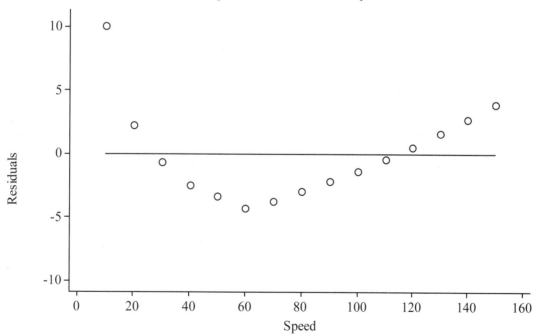

Scatterplot of Residuals versus Speed

Exercise 5.10

(a) Using statistical software, we obtain the following plot.

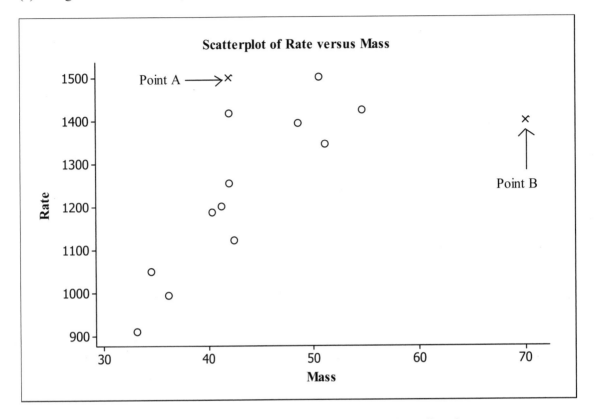

Point A is an outlier in the y direction and point B is an outlier in the x direction.

(b) Using statistical software, we obtain the equations of the three least-squares regression lines.

• Original 12 points

   Least-squares regression line: Rate = 201 + 24.0 × (Mass)

• Original 12 points plus Point A

   Least-squares regression line: Rate = 246 + 23.5 × (Mass)

• Original 12 points plus Point B

   Least-squares regression line: Rate = 620 + 13.9 × (Mass)

We add the least-squares regression lines to the scatterplot.

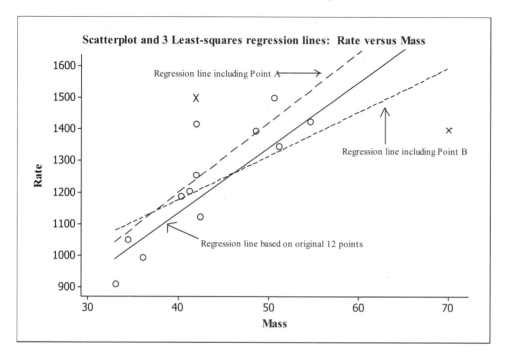

The plot clearly shows that point B is more influential than point A.

Point B is an outlier in the x direction. It lies well to the right of the original 12 points. Visually, it has the effect of making the overall pattern in the plot look flatter. In other words, the overall pattern is best described by a "flatter" line. This explains the smaller slope and larger y intercept.

Point A is an outlier in the y direction. It lies a bit above the original 12 points. Visually, it has the effect of pulling the best fitting line up. This impacts the y intercept more than the slope.

Exercise 5.12

(a) We find that the least-squares regression line is

$$\text{Manatees} = -43.8 + 0.130 \times (\text{Boats})$$

(b) Our prediction for number of manatees killed in 2007, when Boats = 1027:

$$\text{Manatees} = -43.8 + 0.130 \times (1027) = 89.71$$

We would predict 89.71 manatees killed by boats in 2007.

(c) Our prediction for number of manatees killed when there are no boats (Boats = 0):

$$\text{Manatees} = -43.8 + 0.130 \times (0) = -43.8$$

This result is unreasonable because the number of manatees killed can't be less than 0! The problem is, using our least-squares regression line to make this prediction is unreasonable, that Boats = 0 is far outside the range of values of Boats (447 to 1024) upon which the computed line was based. This is an example of extrapolation.

Exercise 5.32

(a) In the problem, the x variable is pre-exam total and the y variable is final-exam score. Given that

$$\bar{x} = 280 \qquad s_x = 30$$
$$\bar{y} = 75 \qquad s_y = 8$$
$$r = 0.6$$

the slope is

$$b = r\frac{s_y}{s_x} = 0.6\frac{8}{30} = 0.16$$

and the intercept is

$$a = \bar{y} - b\bar{x} = 75 - 0.16 \times 280 = 75 - 44.8 = 30.2$$

The equation of the least-squares regression line is therefore

$$\text{final-exam score} = 30.2 + 0.16 \times \text{(pre-exam total)}$$

(b) We use the equation of the least-squares regression line we found in part (a), namely,

$$\text{final-exam score} = 30.2 + 0.16 \times \text{(pre-exam total)}$$

to make our prediction. We calculate (Julie's pre-exam total is 300)

$$\text{predicted final-exam score} = 30.2 + 0.16 \times (300) = 30.2 + 48 = 78.2$$

(c) Recall that the square of the correlation, r^2, is the fraction of the variation in the values of y (in this case final-exam score) that is explained by the least-squares regression of y on x (in this case pre-exam total). Since here $r = 0.6$,

$$r^2 = (0.6)^2 = 0.36$$

Converting this fraction to a percent we find that the observed variation in these students' final-exam scores that can be explained by the linear relationship between final-exam score and pre-exam total is 36%. Thus, only a modest percentage of the observed variation in these students' final-exam scores can be explained by the linear relationship between final-exam score and pre-exam total. In other words, the prediction of Julie's final-exam score based on the equation of the least-squares regression line is probably not very accurate and her score could have been much higher (or much lower) than the predicted value of 78.2.

Exercise 5.42

People who are overweight and trying to lose weight are among those likely to use artificial sweeteners. Thus, the weight of an individual when the person begins to use an artificial sweetener is a lurking variable. We would expect a larger proportion of overweight individuals to be among those using artificial sweeteners than among those using sugar, not because artificial sweeteners cause you to gain weight, but because overweight people are more likely to choose to use artificial sweeteners.

Exercise 5.54

State. The problem asks us to uncover the nature and strength of the effect of decreasing snow cover in the vast land mass of Europe and Asia on summer wind stress. Presumably, higher values of summer wind stress imply stronger monsoons.

Plan. A scatterplot of wind stress (the response) versus snow cover (the explanatory variable) provides information about the nature of the relationship between these variables. The least-squares regression line also provides information about the nature of this relationship. For example, the slope tells us whether decreasing snow cover is associated with increases or decreases in wind stress. Thus, it will be helpful to make both a scatterplot and compute the equation of the least-squares regression line.

The correlation r, or better yet, r^2, provides information about the strength of the relationship. Thus, it will be helpful to calculate both of these quantities.

Solve. Here is a scatterplot of wind stress versus snow cover.

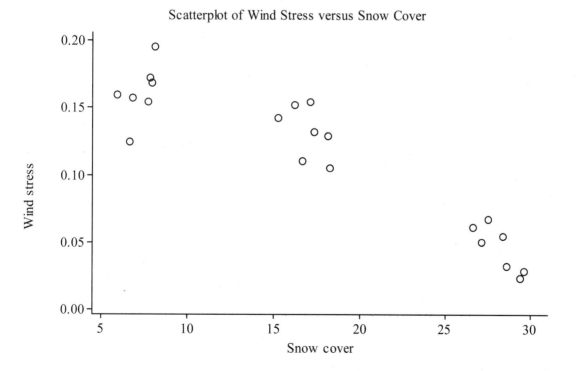

The plot shows an approximately linear trend with wind stress decreasing as snow cover increases.

The equation of the least-squares line computed using statistical software follows.

```
wind stress = 0.212 - 0.00561 × (snow cover)
```

The slope is negative, which is consistent with the trend observed in the scatterplot.

The values of r and r^2 are

$$r = 0.918, r^2 = 0.843$$

These values are reasonably large. In particular, the value of r^2 tells us that 84.3% of the variation in wind stress in the data is explained by the least-squares regression of wind stress on snow cover. We would describe the relation as strong.

Conclude. One must be careful about concluding cause and effect from such a study because it is not a designed experiment. However, the data suggest that decreasing snow cover is associated with greater wind stress. Furthermore, the relationship appears to be fairly strong. The data are certainly consistent with the hypothesis that decreasing snow cover causes an increase in wind stress.

CHAPTER 6

TWO-WAY TABLES

OVERVIEW

This chapter discusses techniques for describing the relationship between two or more categorical variables. To analyze categorical variables, we use counts (frequencies) or percents (relative frequencies) of individuals that fall into various categories. A **two-way table** of such counts is used to organize data about two categorical variables. Values of the **row variable** label the rows that run across the table, and values of the **column variable** label the columns that run down the table. In each cell (intersection of a row and column) of the table, we enter the number of cases for which the row and column variables have the values (categories) corresponding to that cell.

The **row totals** and **column totals** in a **two-way table** give the marginal distributions of the two variables separately. It is usually clearest to present these distributions as percents of the table total. **Marginal distributions** do not give any information about the relationship between the variables. **Bar graphs** are a useful way of presenting these marginal distributions.

The **conditional distributions** in a two-way table help us see relationships between two categorical variables. To find the conditional distribution of the row variable for a specific value of the column variable, look only at that one column in the table. Express each entry in the column as a percent of the column total. There is a conditional distribution of the row variable for each column in the table. Comparing these conditional distributions is one way to describe the association between the row and column variables, particularly if the column variable is the explanatory variable. When the row variable is explanatory, find the conditional distribution of the column variable for each row and compare these distributions. Side-by-side bar graphs of the conditional distributions of the row or column variable can be used to compare these distributions and describe any association that may be present.

Data on three categorical variables can be presented as separate two-way tables for each value of the third variable. An association between two variables that holds for each level of this third variable can be changed, even reversed, when the data are combined by summing over all values of the third variable. **Simpson's paradox** refers to such reversals of an association.

GUIDED SOLUTIONS

Exercise 6.1

KEY CONCEPTS: Marginal distribution

(a) Each of the six entries in the table correspond to a different group of people, and all people in the survey fall into exactly one of these six groups. Add up the six entries to determine how many people these data describe.

(b) The marginal distribution of opinion about quality of recycled filters refers to the opinions of all people, whether they are buyers are non-buyers.

To obtain this marginal distribution, for each level of opinion ("higher," "the same," or "lower"), add the two corresponding table entries. Usually, we give marginal distributions in percents. Convert the number of consumers in each opinion category into the appropriate percentage by dividing by the total number of consumers in the study.

	Higher	The same	Lower
Percent of consumers			

The percent of consumers that think quality of the recycled product is the same or higher than the quality of the other product can be obtained from the marginal distribution above by adding the number of consumers in these opinion categories and dividing by the total number of consumers described by these data.

Exercise 6.3

KEY CONCEPTS: Conditional distribution

We want to compute the conditional distribution of "opinion" for each type of consumer (buyer and non-buyer).

To do this, first determine the number of consumers of each type.

Number of buyers = 20 + 7 + 9 = 36
Number of non-buyers =

Now complete the following table. One entry is provided. Percentages should sum to 100% in each row.

	Opinion of quality		
	Higher	The same	Lower
Buyer	20/36 = 55.6%		
Non-buyer			

Examine this table. Do buyers of recycled filters tend to feel better about their quality than non-buyers do?

Exercise 6.4

KEY CONCEPTS: Conditional distribution, describing relationships, the four-step process

The four-step process involves the following steps:

State. What is the practical question in the context of the real-world setting?

Plan. What specific statistical operations does this problem call for?

Solve. Make the graphs and carry out the calculations needed for this problem.

Conclude. Give your practical conclusion in the setting of the real-world problem.

Here are some suggestions for applying the steps to this problem:

State. What issue regarding gender and college enrollment rates are we trying to investigate?

Plan. What conditional distribution do you need to calculate to address the issue you identified in the *State* step?

Solve. Compute the conditional distributions and describe any patterns you see. To compute the conditional distribution of gender for each age group, first compute the total number of people in each group. To do this, add up the two entries in each column of the table in Exercise 6.2. We have done this for the first age group.

Total number of students aged 15–17 = 116 + 61 = 177
Total number of students aged 18–24 =
Total number of students aged 25–34 =
Total number of students aged 35 or older =

To compute the percent of each age group who are female, divide each entry in the row labeled "Female" by the total number in the group and convert to a percent. We have done the calculation for the first age group. Fill in the rest of the table to complete the exercise.

	% Female	% Male
Age 15–17	116/177 = 65.5%	61/177= 34.5%
Age 18–24		
Age 25–34		
Age 35 or older		

Conclude. Compare the percentage of women in each of the older age groups to the percentage of women in the traditional 18–24 age group. What does this comparison say about our suspicion?

Exercise 6.19

KEY CONCEPTS: Marginal and conditional distributions, describing relationships

(a) There are three treatment groups described: 24 patients received desipramine, 24 received lithium, and 24 received a placebo. Every patient is categorized by treatment received and by whether they relapsed or had no relapse. Fill in the following table. Each entry should be a number (not a percent) of patients. Two of the entries are provided.

	Desipramine	Lithium	Placebo
Relapse			
No relapse	14		
	24		

(b) To compare the effectiveness of the three treatments in preventing relapse, compute the conditional distributions of relapse/no relapse for the three drugs. We have computed the percents for desipramine in the following table. Complete the table by computing the percents for lithium and placebo.

	Desipramine	Lithium	Placebo
Relapse			
No relapse	$(14/24) \times 100\% = 58.33\%$		

Construct the following bar chart for comparing the percent that had no relapse for each treatment. Each treatment should have one bar whose height corresponds to the percentage of patients with no relapse.

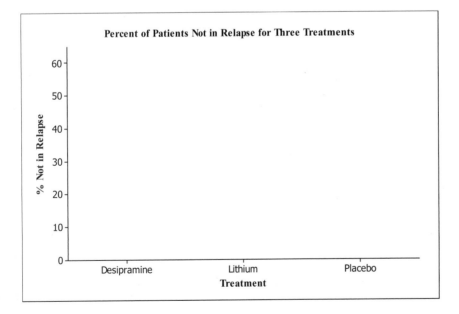

What do you observe?

Exercise 6.25

KEY CONCEPTS: Two-way tables, Simpson's paradox

(a) Add corresponding entries in the two tables and enter the sums in the table.

	Admit	Deny
Male		
Female		

(b) Convert your table in part (a) to one involving percentages of the row totals.

	Admit	Deny
Male		
Female		

(c) Repeat the type of calculations you did in part (b) for each of the original tables.

Business

	Admit	Deny
Male		
Female		

Law

	Admit	Deny
Male		
Female		

(d) To explain the apparent contradiction observed in part (c), consider which professional school is easier to get into and which professional school males and females tend to apply to. Write your answer in plain English in the space provided. Avoid jargon, and be clear!

COMPLETE SOLUTIONS

Exercise 6.1

(a) The number of people these data describe is $20 + 7 + 9 + 29 + 25 + 43 = 133$.

(b) i) There were $20 + 29 = 49$ consumers that felt recycled filters had higher quality. This translates to $49 / 133 = 36.8\%$ of all consumers in the study.
   ii) The percentage of consumers that felt recycled filters have the same quality is $(7 + 25)/133 = 32/133 = 24.1\%$.
   iii) The percentage of consumers that felt recycled filters have lower quality is $(9 + 43)/133 = 52/133 = 39.1\%$.

The marginal distribution (in percents) of consumer opinion of recycled filter quality is summarized in the following table:

	Higher	The same	Lower
Percent	36.8%	24.1%	39.1%

From this distribution, the percentage of consumers that feel that quality of recycled filters is the same or better than that of other filters is $24.1\% + 36.8\% = 60.9\%$.

Exercise 6.3

There are $20 + 7 + 9 = 36$ buyers of recycled filters and $29 + 25 + 43 = 97$ non-buyers.
For each type of consumer, the conditional distribution of opinion on quality of recycled filters (compared with other products) is summarized in the table below.

	Opinion of quality		
	Higher	The same	Lower
Buyer	20/36 = 55.6%	7/36 = 19.4%	9/36 = 25%
Non-buyer	29/97 = 29.9%	25/97 = 25.8%	43/97 = 44.3%

Notice that 55.6% of buyers think highly of the quality of recycled filters, while only 29.9% of non-buyers feel the same way. Similarly, 44.3% of non-buyers feel that recycled filters are of lower quality, while only 25% of buyers feel the same way. It's clear that buyers of recycled filters feel better about their quality than non-buyers.

Exercise 6.4

State. We suspect that a higher percentage of students in older age groups are female.

Plan. We should calculate the conditional distribution of gender for each age group.

Solve.

	% Female	% Male
Age 15–17	116/177 = 65.5%	61/177 = 34.5%
Age 18–24	5470/10161 = 53.8%	4691/10161 = 46.2%
Age 25–34	1319/2143 = 61.5%	824/2143 = 38.5%
Age 35 or older	1075/1691 = 63.5%	616/1691 = 36.5%

Conclude. The data suggest that for each of the older age groups (25–34 and 35 and older), the percentage female is notably higher than in the traditional age 18–24 age group. In other words, according to the data, older students are more heavily made up of females. The data support the hypothesis that the percent of women is higher among older students than in the traditional 18–24 age group.

Exercise 6.19

(a) The complete two-way table is provided:

	Desipramine	Lithium	Placebo
Relapse	10	18	20
No relapse	14	6	4
	24	24	24

(b) The conditional distributions of relapse/no relapse for the three drugs follows:

	Desipramine	Lithium	Placebo
Relapse	10/24= 41.67%	18/24= 75%	20/24= 83.33%
No relapse	14/24 = 58.33%	6/24 = 25%	4/24= 16.67%

Here is a bar graph that displays this information.

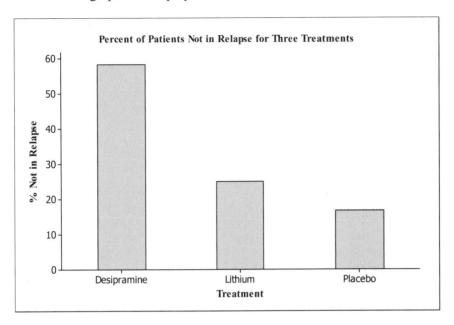

The data show that people taking desipramine had a higher rate of non-relapse (a lower rate of relapse) than people taking either lithium or a placebo.

Exercise 6.25

(a) Here is the desired two-way table:

	Admit	Deny
Male	490	210
Female	280	220

(b) We first add a column containing the row totals to the table in part (a).

	Admit	Deny	Total
Male	490	210	700
Female	280	220	500

We now convert the table entries to percents of the row totals. We divide the entries in the first row by 700 and express the results as percents. We divide the entries in the second row by 500 and express them as percents.

	Admit	Deny
Male	70%	30%
Female	56%	44%

We see that Wabash Tech admits a higher percent of male applicants.

(c) We repeat the calculations in part (b) for each of the original two tables.

Business

	Admit	Deny	Total
Male	480	120	600
Female	180	20	200

Law

	Admit	Deny	Total
Male	10	90	100
Female	100	200	300

Converting entries to percents of the row totals yields the following:

Business

	Admit	Deny
Male	80%	20%
Female	90%	10%

	Admit	Deny
Male	10%	90%
Female	33.3%	66.7%

We see that each school admits a higher percentage of female applicants.

(d) Although both schools admit a higher percentage of female applicants, the admission rates are quite different. Business admits a high percentage of all applicants; it is easier to get into the business school. Law admits a lower percentage of applicants; it is harder to get into the law school. Most of the male applicants to Wabash apply to the business school with its easy admission standards. Thus, overall, a high percentage of males are admitted to Wabash. The majority of female applicants apply to the law school. Because it has tougher admission standards, this makes the overall admission rate of females appear low, even though more females are admitted to both schools!

CHAPTER 7

EXPLORING DATA: PART I REVIEW

To assist you in reviewing the material in Chapters 1–6, we provide the text chapter and related exercises in this Study Guide for each of the odd-numbered review exercises. Other than pointing you in the right direction, we provide no additional hints or solutions. At this point, you should be able to work these exercises on your own with minimal assistance. As a final challenge, we encourage you to work some of the Supplementary Exercises, which integrate more fully the material in these chapters.

Exercise 7.1
Text location – Chapter 1 for types of variables
Related Study Guide exercise – Exercise 1.1

Exercise 7.3
Text location – Chapter 1 for stemplots
Related Study Guide exercise – Exercise 1.11

Exercise 7.5
Text location – Chapter 1 for drawing histograms, Chapter 2 for boxplots
Related Study Guide exercises – Exercises 1.45, 2.29, 2.35

Exercise 7.7
Text location – Chapter 1 for drawing bar graphs and pie charts
Related Study Guide exercise – Exercise 1.25

Exercise 7.9
Text location – Chapter 2 for five-number summary and boxplots
Related Study Guide exercises – Exercises 2.29, 2.35

Exercise 7.11
Text location – Chapter 2 for spotting suspected outliers
Related Study Guide exercise – Exercise 2.35

Exercise 7.13
(a) Text location – Chapter 1 for drawing stemplots
Related Study Guide exercise – Exercise 1.11

(b) Text location – Chapter 2 for mean and median
Related Study Guide exercises – Exercises 2.10, 2.35

Exercise 7.15
Text location – Chapter 3 for 68–95–99.7 rule
Related Study Guide exercise – Exercise 3.7

Exercise 7.17
Text location – Chapter 3 for Normal calculations
Related Study Guide exercise – Exercises 3.43

Exercise 7.19
Text location – Chapter 4 for scatterplots and correlation
Related Study Guide exercises – Exercises 4.8, 4.10

Exercise 7.21
Text location – Chapter 5 for finding the least-squares regression line equation, prediction
and extrapolation
Related Study Guide exercises – Exercises 5.4, 5.12

Exercise 7.23
(a) Text location – Chapter 2 for computing the mean
Related Study Guide exercises – Exercises 2.10, 2.11

(b) Text location – Chapter 4 for adding categorical variables to scatterplots; Chapter 5 for computing
the least-squares regression line equation and adding it to a scatterplot
Related Study Guide exercises – Exercises 4.9, 5.8

Exercise 7.25
Text location – Chapter 4 for scatterplots and correlation
Related Study Guide exercise – Exercise 4.10

Exercise 7.27
(a) Examine the data and count the number of cases in which response to monkey call is stronger than
response to pure tone.

(b) Text location – Chapter 4 for making scatterplots and computing correlation
Related Study Guide exercise – Exercise 4.39

Exercise 7.29
Text location – Chapter 4 for interpreting correlation
Related Study Guide exercise – Exercise 4.37

Exercise 7.31
Text location – Chapter 4 for interpreting correlation; Chapter 5 for interpreting r^2
Related Study Guide exercises – Exercises 4.37, 5.6

Exercise 7.33
(a) Text location – Chapter 5 for computing the least-squares regression line equation, interpreting slope
Related Study Guide exercises – 5.4, 5.32

(b) Text location – Chapter 5 for using the least-squares regression line for making predictions
Related Study Guide exercise – 5.4

CHAPTER 8

PRODUCING DATA: SAMPLING

OVERVIEW

Sampling, when done properly, can yield reliable information about a **population.** The population is the entire group of individuals or objects about which we want information. The information collected is contained in a **sample,** a part of the population observed. How the sample is chosen (the **sampling design**) has a large impact on the usefulness of the data. A useful sample will be representative of the **population** and will help answer our questions. "Good" methods of collecting a sample include the following:

> **Simple random samples,** sometime denoted **SRS**
> **Probability samples**
> **Stratified random samples**
> **Multistage samples**

All these sampling methods involve some aspect of randomness through the use of a formal chance mechanism. Random selection is just one precaution that a person can take to reduce **bias,** the systematic favoring of a certain outcome. Samples we select using our own judgment or because they are convenient are usually biased. Hence, we use computers, calculators, or a **table of random digits** to help us select a sample.

A simple random sample, or SRS, of size n is a collection of n individuals chosen from the population in a manner so that each possible set of n individuals has an equal chance of being selected. In practice, a simple method such as drawing names from a hat is one way of getting an SRS. The names in the hat are the individuals in the population. To choose the sample, we mix up the names and select the sample of n at random from the hat. In reality, the population may be very large and a computer, calculator, or a table of random numbers can be used to as an alternative to drawing the names from the hat.

The method of selecting an SRS using a table of random digits can be summed up in two steps:

1. Give every individual in the population its own numerical label. All labels need to have the same number of digits.
2. Starting anywhere in the table (usually a spot selected at random), read off labels until you have selected as many labels as needed for the sample.

Another common type of sample design is a stratified random sample. First, the population is divided into **strata,** and then an SRS is chosen from each stratum. The strata are formed using some known characteristic of each individual thought to be associated with the response to be measured. Examples of strata are gender or age. Individuals in a particular stratum should be more like one another than those in the other strata.

Poor sample designs include the **voluntary response sample,** where people place themselves in the sample, and the **convenience sample.** Both these methods rely on personal decision for the selection of the sample, generally a guarantee of bias in the selection of the sample.

Other kinds of bias can occur even in well designed studies:

Nonresponse bias, which occurs when individuals who are selected do not participate or cannot be contacted

Bias in the **wording of questions,** leading the answers in a certain direction

Confounding, or confusing, the effect of two or more variables

Undercoverage, which occurs when some group in the population is given either no chance or a much smaller chance than other groups to be in the sample

Response bias, which occurs when individuals participate but are not responding truthfully or accurately due to the way the question is worded, the presence of an observer, fear of a negative reaction from the interviewer, or any other such source

These types of bias can occur even in a randomly chosen sample, and we need to try to reduce their impact as much as possible.

GUIDED SOLUTIONS

Exercise 8.1

KEY CONCEPTS: Population, samples

(a) What population is of interest to the political scientist? What population is the sample from? Are the two populations the same in this case?

(b) How many individuals responded to the question? Those individuals make up the sample.

Exercise 8.7

KEY CONCEPTS: Selecting an SRS with a table of random numbers

The table of random numbers can be used to select an SRS of numbers. To use it to select a random sample of six minority managers, the managers need to be assigned numerical labels. So that everyone does the problem the "same" way, we have labeled the managers according to alphabetical order.

01 Abdulhamid	08 Duncan	15 Huang	22 Puri
02 Agarwal	09 Fernandez	16 Kim	23 Richards
03 Baxter	10 Fleming	17 Lumumba	24 Rodriguez
04 Bonds	11 Gates	18 Mourning	25 Santiago
05 Brown	12 Gomez	19 Nguyen	26 Shen
06 Castro	13 Gupta	20 Peters	27 Vargas
07 Chavez	14 Hernandez	21 Pena	28 Wang

If you go to line 139 in the table and start selecting two-digit numbers, you should get the same answer as the one given in the complete solution. If your entire sample has not been selected by the end of line 139, continue to the next line in the table.

The sample consists of the managers:

Exercise 8.11

KEY CONCEPTS: Stratified random sample

The two strata from which you are going to sample are the suburban townships and the townships that make up Chicago. A stratified random sample consists of combining an SRS from each stratum to form the full sample. How large an SRS will be taken from each of the two strata? How would you label the units in each of the two strata from which you will sample? In the complete solution we have labeled the units within each strata in alphabetic order so you would need to do this as well to get the same sample. Fill in the following table to describe your sampling plan.

Strata

	Suburban townships	Chicago townships
Number of units in stratum		
Sample size		
Labeling method		

Starting on line 101 of Table B, select six of the suburban townships at random using the method in Exercise 8.7.

Suburban townships:

Now, starting on line 110 of Table B, select four of the Chicago townships.

Chicago townships:

The ten suburban and Chicago townships selected are your stratified sample.

Exercise 8.13

KEY CONCEPTS: Nonresponse, response bias

In which of the two periods do you expect the nonresponse to be higher? Is a particular group going to be underrepresented in one of the samples, and could this make the sample less reliable?

Exercise 8.37

KEY CONCEPTS: Sources of bias

If a behavior is socially unacceptable or illegal, do you think greater or fewer people will admit to the behavior in a survey than actually exhibit the behavior. How does this apply to this exercise?

Exercise 8.41

KEY CONCEPTS: Systematic sampling

(a) This exercise is like the example except that there are now 200 addresses instead of 100, and the sample size is now 5 instead of 4. With these two changes, you need to think about how many different systematic samples there are. Two different systematic samples follow:

Systematic sample 1 = 01, 41, 81, 121, 161
Systematic sample 2 = 02, 42, 82, 122, 162

How many systematic samples are there altogether? Choosing one of these systematic samples at random is equivalent to choosing the first address in the sample. The remaining four addresses follow automatically by adding 40. Carry this calculation out using line 120 in the table.

(b) Why are all addresses equally likely to be selected? First, how many systematic samples contain each address? The chance of selecting an address is the same as the chance of selecting the systematic sample that contains it. With this in mind, what is the chance of any address being chosen? The definition of an SRS requires that all samples of five addresses are equally likely to be selected. In a systematic sample, are all samples of five addresses even possible?

Exercise 8.42

KEY CONCEPTS: Sampling frame, undercoverage

(a) Which households wouldn't be in the sampling frame? Make some educated guesses as to how these households might differ from those in the sampling frame (other than the fact that they don't have a phone number in the directory).

(b) Random digit dialing makes the sampling frame larger. Which households are added to it?

Exercise 8.44

KEY CONCEPTS: Wording of questions

A question can be worded in a way to make it seem as though any reasonable person should answer yes (or no). Which questions are slanted toward a desired response? Are all questions clear?

(a)

(b)

(c)

Exercise 8.46

KEY CONCEPTS: Populations, samples, sample size, bias

(a) What is the population of interest and the sample? Is the sample from the population of interest, or are there potential sources of bias?

(b) Is this a scientific sample using probability in the selection of the sample? Is the sample size large?

COMPLETE SOLUTIONS

Exercise 8.1

(a) The population of interest to the political scientist is all college students. Unfortunately, for convenience, the sample is selected from only the undergraduates attending her college. Suppose her college is a small liberal eastern college or a conservative evangelical college. In these cases, the opinions of students attending her college on a political issue such as social security may differ from opinions of the population of all college students, biasing the results.

(b) The sample is the 104 questionnaires returned. More than 50% of the 250 students to whom the questionnaire was mailed did not respond, which can introduce nonresponse bias.

Exercise 8.7

To choose an SRS of six managers to be interviewed, first label the members of the population by associating a two-digit number with each.

01 Abdulhamid	08 Duncan	15 Huang	22 Puri
02 Agarwal	09 Fernandez	16 Kim	23 Richards
03 Baxter	10 Fleming	17 Lumumba	24 Rodriguez
04 Bonds	11 Gates	18 Mourning	25 Santiago
05 Brown	12 Gomez	19 Nguyen	26 Shen
06 Castro	13 Gupta	20 Peters	27 Vargas
07 Chavez	14 Hernandez	21 Pena	28 Wang

Now, go to Table B in your textbook and read two-digit groups until six managers are chosen. Starting at line 139,

55|58|8 9|94|04| 70|70|8 4|10|98 43|56|3 5|69|34| 48|39|4 5|17|19| 12|97|5 1|32|58| 13|04|8

The selected sample is 04 Bonds, 10 Fleming, 17 Lumumba, 19 Nguyen, 12 Gomez, and 13 Gupta.

Exercise 8.11

The table below describes the sampling plan:

	Strata	
	Suburban townships	Chicago townships
Number of units in stratum	30	8
Sample size	6	4
Labeling method	Alphabetic order	Alphabetic order

First, label 30 suburban townships by associating a two-digit number with each.

01 Barrington	07 Elk Grove	13 Maine	19 Orland	25 Riverside
02 Berwyn	08 Evanston	14 New Trier	20 Palatine	26 Schaumburg
03 Bloom	09 Hanover	15 Niles	21 Palos	27 Stickney
04 Bremen	10 Lemont	16 Northfield	22 Proviso	28 Thornton
05 Calamet	11 Leyden	17 Norwood Park	23 Rich	29 Wheeling
06 Cicero	12 Lyons	18 Oak Park	24 River Forest	30 Worth

Go to Table B in your textbook and read two-digit groups until eight townships are chosen. Starting at line 101,

<u>19</u>|22|3 9|50|34| <u>05</u>|75|6 2|87|<u>13</u> 96|40|9 1|<u>25</u>|31| 42|54|4 8|<u>28</u>|53|

Suburban townships: Orland, Proviso, Calamet, Maine, Riverside, and Thornton

Now, label the eight Chicago townships by associating a one-digit number with each.

1 Hyde Park	3 Lake	5 North Chicago	7 South Chicago
2 Jefferson	4 Lake View	6 Rogers Park	8 West Chicago

Go to Table B in your textbook and read one-digit numbers until 4 townships are chosen. Starting at line 110,

<u>3</u> <u>8</u> <u>4</u> 4 8   4 8 <u>7</u> 8 9

Chicago townships: Lake, West Chicago, Lake View, and South Chicago

The ten suburban and Chicago townships selected in this manner are your stratified sample.

Exercise 8.13

You would expect that the higher rate of no-answer was probably during the second period as more families are likely to be gone for vacation. Nonresponse can always bias the results. In this case, those who are more affluent may be more likely to travel during the summer months and they would be underrepresented in the sample. Their views could be different, which would bias the results.

Exercise 8.37

If a behavior is socially unacceptable, embarrassing, or illegal, many people will not admit to it in a survey. In this case, some people who don't wear seatbelts are embarrassed to admit it, which makes the percentage in the survey who claim to be wearing seatbelts higher than the number that are actually wearing them. Not wearing seatbelts is illegal as well.

Exercise 8.41

(a) We want to select 5 addresses out of 200, so we think of the 200 addresses as 40 lists, each containing 5 addresses. We choose one address from the first 40 and then every 40th address after that. The first step is to go to Table B, line 120, and choose the first two-digit random number you encounter that is one of the numbers 01, $\cdots$, 40.

<u>35</u>476

The selected number is 35, so the sample includes addresses numbered 35, 75, 115, 155, and 195.

(b) Each individual is in exactly one systematic sample, and the systematic samples are equally likely to be chosen. In our previous example, there were 40 systematic samples, each containing 5 addresses. The chance of selecting any address is the chance of picking the systematic sample that contains it, which is 1 in 40.

A simple random sample of size n would allow every set of n individuals an equal chance of being selected. Thus, in this exercise, when using an SRS the sample consisting of the addresses numbered 1, 2, 3, 4, and 5 would have the same probability of being selected as any other set of 5 addresses. For a systematically selected sample, all samples of size n do not have the same probability of being selected. The sample consisting of the addresses numbered 1, 2, 3, 4, and 5 would have zero chance of being selected since the numbers of the addresses do not all differ by 40. The sample we selected in (a), 35, 75, 115, 155, and 195, had a 1 in 40 chance of being selected, so all samples of 5 addresses are not equally likely.

Exercise 8.42

(a) Households omitted from the frame are those that do not have a telephone number listed in the telephone directory. The types of people who might be underrepresented are poorer people (including the homeless) who cannot afford a phone and the group of people who have unlisted numbers. It is harder to characterize this second group. As a group, they would tend to have more money, as you need to pay to have your phone number unlisted, or the group might include more single women who do not want their phone numbers available and possibly people whose jobs put them in contact with large groups of people who might harass them if their phone numbers were easily accessible.

(b) People with unlisted numbers will be included in the sampling frame. The sampling frame would now include any household with a phone. One interesting point is that all households would not have the same probability of being in the sample as some households have multiple phone lines and would be more likely to get into the sample. So, strictly speaking, random-digit dialing would actually provide not an SRS of households with phones but an SRS of phone numbers!

Exercise 8.44

(a) The beginning of the question suggests that cell phone use is associated with brain cancer. This initial suggestion and the wording "the danger of using cell phones" would lead most reasonable people to be in favor of including a warning label. The question is slanted in favor of this response.

(b) The question is clear but is slanted in favor of national health insurance. The reason for agreeing with a question should not be contained within the question.

(c) The question is slanted because it contains reasons for not supporting government subsidies for day care programs. As a question, the wording is a little technical and unclear, using phrases such as "negative externalities in parent labor force participation" and "increased group size with morbidity of children." A simpler, unslanted version such as "Do you support government subsidies for day care programs?" would be better.

Exercise 8.46

(a) The population is all people who live in Ontario. Because everyone uses health care, it is not restricted to adults, and so on. The sample is the 61,239 residents of Ontario who were interviewed.

(b) Yes. This is a very large sample and it is a probability sample, so we expect that the sample proportions are quite close to the population proportions.

CHAPTER 9

PRODUCING DATA: EXPERIMENTS

OVERVIEW

Data can be produced in a variety of ways. **Observational studies** are investigations in which the state of some population is simply observed, usually with data collected by sampling. Even with proper sampling, data from observational studies are generally not appropriate for investigating cause-and-effect relations between variables. The reason is that the explanatory variable can be **confounded** with lurking variables, so its effects on the response cannot be distinguished from those of the lurking variables. **Experiments** are investigations in which data are generated by active imposition of some treatment on the subjects of the experiment. Properly designed experiments are the best way to investigate cause-and-effect relations between variables.

In an **experiment,** one or more **treatments** are imposed on experimental **units** or **subjects.** A treatment is a combination of **levels** of the explanatory variables, called **factors.** The design of an experiment is a specification of the treatments to be used and the manner in which units or subjects are assigned to these treatments. The basic features of well-designed experiments are **control, randomization,** and **replication.**

Control is used to avoid confounding (mixing up) the effects of treatments with other influences such as lurking variables. One such lurking variable is the **placebo effect,** which is the response of a subject to the fact of receiving any treatment. The simplest form of control is **randomized comparative experimentation,** which involves comparisons between two or more treatments. One of these treatments may be a **placebo** (fake treatment), and subjects receiving the placebo are referred to as a **control group.**

Randomization can be carried out using the ideas in Chapter 8 of your text. Randomization is carried out before applying the treatments and helps control bias by creating treatment groups that are similar. **Replication,** the use of many units in an experiment, is important because it reduces the chance variation between treatment groups arising from randomization. Using more units helps increase the ability of your experiment to establish differences between treatments.

Further control in an experiment can be achieved by forming experimental units into **blocks** that are similar in some way thought to affect the response, similar to strata in a stratified sample design. In a **block design,** units are first formed into blocks, and then randomization is carried out separately in each block. **Matched pairs** are a simple form of blocking used to compare two treatments. In a matched pairs experiment, either the same unit (the block) receives both treatments in a random order or very similar units are matched in pairs (the blocks). In the latter case, one member of the pair receives one of the treatments and the other member the remaining treatment. Members of a matched pair are assigned to treatments using randomization.

Some additional problems that can occur that are unique to experimental designs are **lack of blinding** and **lack of realism.** These problems should be addressed when designing the experiment.

GUIDED SOLUTIONS

Exercise 9.1

KEY CONCEPTS: Explanatory and response variables, experiments, and observational studies

What are the researchers trying to demonstrate with this study? What groups are being compared, and did the experiment deliberately impose membership in the groups on the subjects to observe their responses? What are the explanatory and response variables?

Explanatory variable:

Response variable:

Observational study or experiment (circle one)? Why?

Exercise 9.5

KEY CONCEPTS: Identifying experimental units, factors, treatments, and response variables

Read the description of the study carefully. To identify the "individuals," ask yourself, what the treatments are going to be applied to. What is being measured on these individuals? This is the response.

There is only one factor in this experiment. How many levels does it have? The levels of the factor are the treatments.

Exercise 9.9

KEY CONCEPTS: Completely randomized design, randomization

(a) How many treatment groups are included in the experiment and how many plots are assigned to each group? This information must be included in your outline. Now, using Figure 9.4 of your text as a model, draw an appropriately labeled diagram that outlines a completely randomized design for the study.

Label the plots 01 to 18. At line 142 of Table B, select an SRS of six plots to receive the control treatment. Continue in Table B, selecting six more to receive added water in winter. The remaining 6 receive added water in spring.

Six plots assigned to receive control treatment:

Six plots assigned to receive added water in winter:

Six plots assigned to receive added water in spring:

Exercise 9.12

KEY CONCEPTS: Randomized comparative experiments, observational studies

Is the new design suggested by the executive an experiment? What are the disadvantages of the approach?

Exercise 9.15

KEY CONCEPTS: Double-blind experiments, bias

What does it mean for the ratings to be blind? Was this done in this case? If not, how can this bias the results and in which direction?

Exercise 9.16

KEY CONCEPTS: Matched pairs design, randomization

First identify the treatments and the response variable. Next, decide what makes up the matched pairs in this experiment. How will you use a coin flip to assign members of a pair to the treatments? What will you measure and how will you decide whether the right-hand tends to be stronger in right-handed people?

Exercise 9.37

KEY CONCEPTS: Design of an experiment, randomization

(a) To begin, identify the subjects, the factors, the treatments, and the response variable.

• How many treatments are there? Hence, how many groups of subjects must you form?

• How will you assign subjects to treatment groups?

• What are the treatments? What will each subject be required to do?

• What response will you measure and how will you decide whether the treatments differ in their effect?

First use a diagram like Figure 9.2 to display the treatments in a design with two factors.

Now outline your design in words or in a drawing similar to Figures 9.3 and 9.4.

(b) Following is the list of names. Assign a numerical label to each one. Be sure to use the same number of digits for each label. So that everyone does the problem in the "same" way, we used the convention of starting with the label 00 for Abbott and continuing to label down the columns (01 Abdalla, etc). Of course, one could start with another number (such as 01) and label across rows if one wished.

Abbott	Decker	Herrera	Lucero	Richter
Abdalla	Devlin	Hersch	Masters	Riley
Alawi	Engel	Hurwitz	Morgan	Samuels
Broden	Fuentes	Irwin	Nelson	Smith
Chai	Garrett	Jiang	Nho	Suarez
Chuang	Gill	Kelley	Ortiz	Upasani
Cordoba	Glover	Kim	Ramdas	Wilson
Custer	Hammond	Landers	Reed	Xiang

Now, start at line 130 in Table B. Read across the row in groups of digits equal to the number of digits you used for your labels (if you used two digits for labels, read line 130 in pairs of digits). Keep reading until you have selected all the names for the first treatment. You may need to continue on to line 131, line 132, and subsequent lines. After you have selected the names for treatment 1, continue in Table B to assign the ten people to receive treatment 2 and then ten to receive treatment 3. The remaining names are assigned to treatment 4.

Exercise 9.43

KEY CONCEPTS: Matched pairs design, randomization

First, identify the subjects, the treatments, and the response variable. Next, decide what makes up the matched pairs in this experiment. How will you use the table of random numbers to assign members of a pair to the treatments? Why is it important to have each player's trials on different days?

The first 25 digits of Table B at line 130 are reproduced here. Use them to decide which players will get oxygen on their first trial.

69051   64817   87174   09517   84534

COMPLETE SOLUTIONS

Exercise 9.1

The explanatory variable is whether or not the person made use of handheld cellular phones (we assume on a regular basis), and the response is whether or not the person contracted brain cancer. No attempt was made to decide which individuals were going to make use of a cellular phone (the treatment), so this is an observational study, not an experiment.

Exercise 9.5

The "subjects" in this experiment are the pine seedlings that we are going to plant. The observed response is the dry weight of the young trees at the end of the experiment.

There is one factor in the experiment, which is the amount of sunlight. This factor is at three levels: full light, 25% light, or 5% light. These three levels are the three treatments in the experiment.

Exercise 9.9

There are three treatments: a control, added water in winter, and added water in spring. There are 18 plots to be used for the study, so 6 plots are assigned at random to each of the treatment groups.

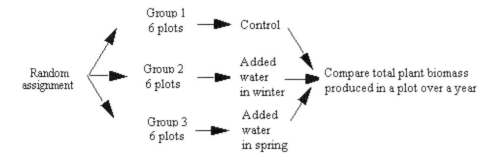

Table B starting at row 142 is reproduced here.   Read across the row in groups of two digits since you used two digits for your labels, skipping repeats. The first six underlined numbers correspond to the plots receiving the control treatment and the next six to those receiving added water in the winter.   The remaining plots receive added water in the spring.

```
72829  50232  97892  63408  77919  44575  24870  04178
88565  42628  17797  49376  61762  16953  88604  12724
62964  88145  83083  69453  46109  59505  69680  00900
19687  12633  57857  95806  09931  02150  43163  58636
37609  59057  66967  83401  60705  02384  90597  93600
54973  86278  88737  74351  47500  84552  19909  67181
00694  05977  19664  65441  20903  62371  22725  53340
71546  05233  53946  68743  72460  27601  45403  88692
07511  88915  41267
```

Plots assigned to receive control treatment: 02, 08, 17, 10, 05, and 09

Plots assigned to receive added water in winter: 06, 16, 01, 07, 18, and 15

Plots assigned to receive added water in spring: 03, 04, 11, 12, 13, and 14

Exercise 9.12

The new design is an observational study, not an experiment. The electric company wants the only systematic differences in groups to be the treatments. Electric use varies from year to year depending on the weather. If charts or indicators are introduced in the second year and the electric consumption in the first year is compared with the second year, you won't know if the observed differences are due to the introduction of the chart or to lurking variables. For example, if the comparison is being made in the summer months, it is possible that the second year had a cooler summer, which reduced the need for air conditioning and reduced electric consumption, rather than the introduction of charts or indicators. A control group ensures that influences other than the introduction of the indicators or charts operate equally on all groups.

Exercise 9.15

The ratings were not blind because the experimenter who rated the subjects' level of anxiety presumably knew whether the subjects were in the meditation group or not. Since the experimenter was hoping to show that those who meditated had lower levels of anxiety, he might unintentionally rate those in the meditation group as having lower levels of anxiety if there is any subjectivity in the ratings. It would be better if a third party who did not know which group the subjects belonged to rated the subjects' anxiety levels.

Exercise 9.16

We have ten subjects. There are two treatments in the study. Treatment 1 is squeezing with the right hand, and treatment 2 is squeezing with the left hand. The response is the force exerted as indicated by the reading on the scale.

To do the experiment, we use a matched pairs design. The matched pairs are the two hands of the subjects. We randomly decide which hand to use first, perhaps by flipping a coin. We measure the response for each hand and then compare the forces for the left and right hands over all subjects to see if there is a systematic difference between the two hands.

Exercise 9.37

In this case, the subjects are the 40 headache sufferers who have agreed to participate in the study. The two factors are antidepressant (placebo or antidepressant given) and stress management training (given or not given), and they form the four treatments:

Treatment 1: Antidepressant and no stress management training

Treatment 2: Placebo (no antidepressant) and no stress management training

Treatment 3: Placebo (no antidepressant) and stress management training

Treatment 4: Antidepressant and stress management training

The response variables are the number of headaches over the study period and some measure of the severity of these headaches. The problem does not specify how the severity is to be measured.

Subjects should be randomly assigned to treatments, with ten assigned to treatment 1, ten assigned to treatment 2, ten assigned to treatment 3, and the remainder assigned to treatment 4. Each subject follows his or her treatment regimen over the course of the study. The average number of headaches and the severity of the headaches for each treatment should be calculated and the results for the four groups compared.

The treatments are displayed in the following diagram.

Factor B
Stress Management

		Yes	No
Factor A Antidepressant	Yes	1	2
	No	3	4

A drawing summarizing the experimental design follows.

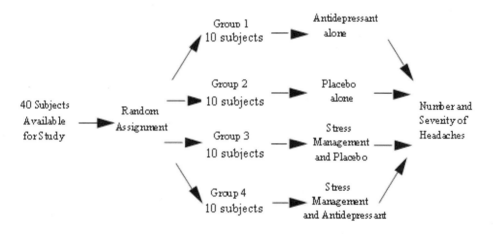

Although you can use the applet or Table B, we illustrate the use of Table B to carry out the random assignment of the subjects to the treatments. First, label the 40 names using two-digit labels. We use the convention of starting with the label 00 and labeling down the columns in alphabetical order. Of course, if one wished, one could start with another number (such as 01) and label across rows.

00 Abbott	08 Decker	16 Herrera	24 Lucero	32 Richter
01 Abdalla	09 Devlin	17 Hersch	25 Masters	33 Riley
02 Alawi	10 Engel	18 Hurwitz	26 Morgan	34 Samuels
03 Broden	11 Fuentes	19 Irwin	27 Nelson	35 Smith
04 Chai	12 Garrett	20 Jiang	28 Nho	36 Suarez
05 Chuang	13 Gill	21 Kelley	29 Ortiz	37 Upasani
06 Cordoba	14 Glover	22 Kim	30 Ramdas	38 Wilson
07 Custer	15 Hammond	23 Landers	31 Reed	39 Xiang

Line 130 from Table B follows. We should read line 130 in pairs of digits from left to right. Vertical bars are placed between consecutive pairs to indicate how to read the table. The pairs that correspond to labels in the list and that have not been previously selected are underlined:

69|<u>05</u>|<u>16</u>|48|<u>17</u>|87|17|40|95|17|84|53|40|64|89|87|<u>20</u>|<u>19</u>|72|45

We find only five of our labels, so we need to continue reading on line 131:

05|<u>00</u>|71|66|<u>32</u>|81|19|41|48|73|<u>04</u>|19|78|55|76|45|19|59|65|65

We now have eight labels, and continuing to line 132 gives us the remaining labels for the group assigned to treatment 1:

68|73|<u>25</u>|52|59|84|<u>29</u>|20|87|96|43|16|59|37|39|31|68|59|71|50

The treatment 1 group consists of Chuang, Herrera, Hersch, Jiang, Irwin, Abbott, Richter, Chai, Masters, and Ortiz. Continuing in line 132 (and ignoring pairs corresponding to previously selected subjects), we assign the next nine subjects to treatment 2:

|52|59|84|29|20|87|96|43|16|59|37|39|<u>31</u>|68|59|71|50

45|74|04|<u>18</u>|<u>07</u>|65|56|<u>13</u>|<u>33</u>|<u>02</u>|07|05|19|36|<u>23</u>|18|13|20|95|47

27 |81|67|84|16|18|32|92|13|37| <u>35</u>|<u>21</u>|33|77|41|04|31|<u>26</u>|85|08

giving the treatment 2 group as Reed, Hurwitz, Custer, Gill, Riley, Alawi, Landers, Smith, Kelley, and Morgan. Continuing in line 134,

27 |81|67|84|16|18|32|92|13|37| 35|21|33|77|41|04|31|26|85|<u>08</u>

66|92|55|56|58|39|<u>10</u>|07|84|58|<u>11</u>|20|61|98|76|87|<u>15</u>|13|<u>12</u>|60

08|42|<u>14</u>|47|53|77|<u>37</u>|72|87|44|75|59|20|85|63|79|14|<u>09</u>|<u>24</u>|54

53|64|56|68|12|61|42|14|78|<u>36</u>

The treatment 3 group is Decker, Engel, Fuentes, Hammond, Garrett, Glover, Upasani, Devlin, Lucero, and Suarez. The ten remaining subjects, Abdilla, Broden, Cordoba, Kim, Nelson, Nho, Ramdas, Samuels, Wilson, and Xiang, are assigned to treatment 4. Using Table B can become tedious with a large number of subjects, so it is best to leave such calculations to a computer.

Exercise 9.43

We have 25 players available and there are two treatments. Treatment 1 is inhaling oxygen during the rest periods, and treatment 2 is not inhaling oxygen between the rest periods. The response is the time on the final run.

To do the experiment, we use a matched pairs design. Each player is a block and each player will run 100 yards four times under both treatments, three in quick succession followed by a fourth time after a 3-minute rest. We should randomly decide which treatment to use first. For the players to have a chance to recover, we should allow sufficient time between the two trials (say, a day or two).

To assign which treatment comes first for each player, suppose the players are numbered from 1 to 25. Go to Table B, line 130, to decide which players get oxygen on their first trial. Use the following randomization. If the random digit is 0, 1, 2, 3, or 4, then the subject gets oxygen on the first trial. Otherwise, the subject gets oxygen on the second trial. The first 25 digits in line 130 follow, with the digits 0, 1, 2, 3, and 4 underlined:

69<u>0</u>51 64<u>8</u>17 87<u>1</u>74 <u>0</u>95<u>1</u>7 84<u>5</u>34

Since positions 1, 3, 7, 9, 13, 15, 16, 19, 22, 24, and 25 correspond to digits 0, 1, 2, 3, or 4, players numbered 1, 3, 7, 9, 13, 15, 16, 19, 22, 24, and 25 receive oxygen during the rest periods on their first trial and no oxygen during their second trial. The remaining players get oxygen during the rest periods on their second trial and none during their first trial. In this example, 11 players get oxygen during the rest periods on their first trial and 14 get oxygen during the rest period on their second trial. When using a table of random numbers or software to generate random numbers, there is no guarantee that the groups will balance out exactly every time the randomization is done.

As an alternative, you could label the players 01 through 25. Begin on line 130 and select 12 players at random to be in the oxygen group. The remaining 13 would not get oxygen during the rest period. This would guarantee balance, or as close to it as you can get with an odd number of players.

CHAPTER 10

INTRODUCING PROBABILITY

OVERVIEW

Probability is the mathematical study of random processes over a long series of repetitions. A process or phenomenon is called **random** if its outcome is uncertain. Although individual outcomes are uncertain, when the process is repeated a large number of times, the underlying distribution for the possible outcomes begins to emerge. For any outcome, its **probability** is the proportion of times, or the relative frequency, with which the outcome would occur in a long series of repetitions of the process. It is important that these repetitions or trials be **independent** for this property to hold. By independent we mean that the outcome of one trial must not influence the outcome of any other.

You can study random behavior by carrying out physical experiments such as coin tossing or die rolling, or you can simulate a random phenomenon on the computer. Using the computer is particularly helpful in considering a large number of trials. Randomness and independence are the keys to using the rules of probability.

The description of a random phenomenon begins with the **sample space, *S*,** which is the list of all possible outcomes. A set of outcomes is called an **event.** Once we have determined the sample space, a **probability model** tells us how to assign probabilities to the various events that can occur. There are four basic rules that probabilities must satisfy:

- Any probability is a number between 0 and 1. $P(A)$ means "the probability of *A*." If the probability is 0, the event will never occur. If the probability is 1, the event will always occur.

- All possible outcomes together must have probability 1.

- The probability that an event does not occur is 1 minus the probability that the event occurs. Using notation,

$$P(A) = 1 - P(\text{not } A)$$

- If two events have no outcomes in common, the probability that one or the other occurs is the sum of their individual probabilities. These events are **disjoint.** This is the addition rule for disjoint events, namely,

$$P(A \text{ or } B) = P(A) + P(B)$$

In a sample space with a finite number of outcomes, probabilities are assigned to the individual outcomes and the probability of any event is the sum of the probabilities of the outcomes it contains. All outcomes must have a probability between 0 and 1; the sum of all probabilities must add up to 1.

Even a sample space with an infinite number of outcomes can have probabilities assigned to it. To assign probabilities, we use a density curve. (Refer to Chapter 3 of your textbook to refresh your memory.) The area under a density curve must be equal to 1. Probabilities are assigned to events as areas under the density curve. The normal distribution is the most useful of the density curves. **Normal distributions are probability models.**

A **random variable** is a variable whose value is a numerical outcome of a random phenomenon. The **probability distribution** of a random variable tells us about the possible values of the random variable and how to assign probabilities to these values. A random variable can be **discrete** or **continuous.** A **discrete random variable** has finitely many possible values. Its distribution gives the probability of each value. A **continuous random variable** takes all values in some interval of numbers. A **density curve** describes the distribution of a continuous random variable.

GUIDED SOLUTIONS

Exercise 10.1

KEY CONCEPTS: Probability as the proportion of times an event occurs in many repeated trials

The probability of any outcome of a random phenomenon is the proportion of times the outcome would occur in a very long series of repetitions. By "very large number of repetitions," we mean many more than 1000. It might help to examine Figure 10.1: In that figure, did we have exactly 500 heads in 1000 coin tosses?

Now, if we play 1000 hands of Texas hold'em, is it likely that we'll have exactly 88 hands in which you hold four of a kind? Why not?

Exercise 10.5

KEY CONCEPTS: Sample space

One of the main difficulties encountered when describing the sample space is finding some notation to express ideas formally. Following the text, our general format is $S = \{\quad\}$, where a description of the outcomes in the sample space is included within the braces.

(a) There are only two possible outcomes to the question "Is the student male or female?" (and they aren't "yes" and "no"!). You would write
$$S = \{\text{female, male}\}$$

(b) It isn't clear what the largest or smallest possible student heights are. For example, if you say that the largest height possible is 90 inches (7' 6''), why not 91 inches? The point is there is no obvious upper bound on student height. Obviously, all student heights must be at least 0 inches (!). Assuming heights are rounded to the nearest whole inch, one possible sample space is

$$S = \{0 \text{ inches, 1 inch, 2 inches, 3 inches, } \ldots\}$$

and it's understood that we'll never actually see a student (say) 4 inches tall.

(c) Obviously, the smallest amount of money in coins a student is carrying is $0.00. Any 1–cent increment is possible. Hence, $0.00, $0.01, $0.02,... . Referring to the problem in (b), is there an upper bound on this set?

$$S =$$

(d) What are the possible responses a student can make? Ignore the possibility of a response like "I don't know." Write S in the space provided. You might assume that +/– grade designations are possible, or you might decide that they aren't. This varies by school.

$$S =$$

Exercise 10.36

KEY CONCEPTS: Probabilities in a finite sample space, legitimate probability models

(a) In a legitimate probability model, the probabilities of all possible outcomes in the sample space must add up to 1. What do the six probabilities listed in this table sum to? What proportion of cars are not accounted for in this table? These cars have some different color.

Probability that the vehicle you choose has a color other than the six listed =

(b) The events "The randomly chosen car is silver" and "the randomly chosen car is white" are disjoint. What do the rules of probability tell us about the probability that the randomly chosen car is either silver or white? Using this information, what do the rules of probability tell us about the probability that the event "The randomly chosen car is either silver or white" does not occur?

Exercise 10.46

KEY CONCEPTS: Sample spaces for simple random sampling, probabilities of events

(a) The sample space needs to include all pairs of two employees without regard to order. Make the list: $S = \{(\text{Abby, Deborah}), (\text{Abby, Mei-Ling}), \text{etc.}\}$. There is no need to include both (Abby, Deborah) and (Deborah, Abby) in your list because each of these mean that Deborah and Abby go to Paris - they're the same event. In other words, order doesn't matter. Write down list.

(b) How many outcomes are there in the sample space in part (a)? If they are equally likely, what is the probability of each?

(c) How many outcomes in S include Mei-Ling? When the outcomes are equally likely, the probability of the event is just

$$\frac{\text{number of outcomes in the event}}{\text{number of outcomes in } S} =$$

(d) How many outcomes in S include neither of the two men?

Exercise 10.49

KEY CONCEPTS: Random numbers, density curve

(a) Recall that a discrete random variable has finitely many possible values. A continuous random variable takes all values in some interval of numbers. Which is true in this example?

(b) The total area underneath any density curve is 1. Since the density curve has constant height over the range 0 to 2, what must the height be to make the area under this curve equal to 1? Note that since the area under the density curve is a rectangular region, the formula for the area of a rectangle will be useful.

Now draw a graph of the density curve.

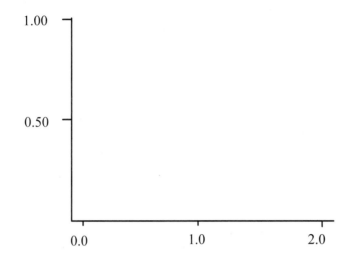

(c) $P(Y \leq 1)$ is the area under the density curve below 1. You may find it helpful to sketch this region on your graph in part (b). This is a rectangular region, so use the formula for the area of a rectangle to compute the area of this region.

Exercise 10.51

KEY CONCEPTS: Normal random variables

(a) V denotes the proportion of the sample who vote, and has the normal distribution with mean $\mu = .56$ and standard deviation $\sigma = .019$. First, draw a graph (picture) of this normal density curve. The event of interest is "V is between .52 and .60," or $\{.52 \leq V \leq .60\}$. Shade in the area corresponding to the probability of this event. This is the sort of problem you worked in Chapter 3. You'll need to convert .52 and .60 into z-scores and use the standard normal distribution table to compute the corresponding probability.

(b) As in (a), shade the region under the normal density curve corresponding to the event $\{V \geq .72\}$. Again, use the methods described in Chapter 3 to compute this area.

Exercise 10.54

KEY CONCEPTS: Estimating probabilities as the proportion of times an event occurs in many repeated trials

Record the number of heads for your 50 spins.

The probability of an event is the proportion of times the event occurs in many repeated trials of a random phenomenon. To estimate the probability of heads when you spin a nickel, determine the proportion of times heads occurred in your 50 trials.

Estimate =

Exercise 10.56

KEY CONCEPTS: Simulating a random phenomenon

(a) You will need to use the applet or software to simulate the 100 trials. As the problem suggests, the key phrase to look for in your software is "Bernoulli trials." Your software may allow you to actually generate the words "Hit" and "Miss" or perhaps the letters H and M. However, your software is more likely to allow you to generate only the numbers 0 and 1. In this case, count a 0 as a miss and a 1 as a hit.

After you do so, the computer can be used to calculate the proportion of hits.

Proportion of hits =

For most students, the proportion of hits will be within 0.05 or 0.10 of the true probability of 0.5.

(b) You need to go through your sequence to determine the longest string of hits or misses.

Longest run of shots hit =   Longest run of shots missed =

COMPLETE SOLUTIONS

Exercise 10.1

To say that the probability of getting four of a kind is 88/1000 means that in a very large number of hands of Texas hold'em in which you hold a pair in your hand, the proportion that will give you four of a kind is 88/1000.

The key phrase here is "a very large number of hands." "Very large" means much, much larger than 1000. Thus, a probability of 88/1000 means the proportion after millions and millions of hands in which you hold a pair, not after 1000.

Exercise 10.5

(a) $S = \{$female, male$\}$
(b) $S = \{$0 inches, 1 inch, 2 inches, 3 inches, . . .$\}$
(c) $S = \{\$0, \$0.01, \$0.02, \$0.03, ...\}$
(d) If grades can have +/– designations, then perhaps $S = \{$F, D–, D, D+, C–, C, C+, B–, B, B+, A–, A$\}$.

Exercise 10.36

(a) The sum of the six probabilities listed in the table = 0.90.

The sum of all possible outcomes must be 1, so the probability that the vehicle you choose has a color other than the six listed when added to 0.90 must make the sum 1. Thus,

Probability the vehicle you choose has a color other than the six listed = $1 - 0.90 = 0.10$

(b) Because the events are disjoint, we simply add the probabilities of silver and white to get

Probability that a randomly chosen vehicle is either silver or white = $0.18 + 0.19 = 0.37$

By probability basic property 4,

P(car is neither silver nor white) = $1 - P$(car is silver or white) = $1 - 0.37 = 0.63$.

Exercise 10.46

(a) S = {(Abby, Deborah), (Abby, Mei-Ling), (Abby, Sam), (Abby, Roberto), (Deborah, Mei-Ling), (Deborah, Sam), (Deborah, Roberto), (Mei-Ling, Sam), (Mei-Ling, Roberto), (Sam, Roberto)}.

(b) There are ten possible outcomes. Since they are equally likely, each has probability 0.10.

(c) Mei-Ling is in four of the outcomes, so her chance of attending the conference in Paris is $4/10 = 0.4$.

(d) The chosen group must contain two women. The elements of S that consist of two women are (Abby, Deborah), (Abby, Mei-Ling), and (Deborah, Mei-Ling). There are three possibilities, so the desired probability is $3/10 = 0.3$.

Exercise 10.49

(a) Here, Y takes on all values in the interval 0 to 2. Thus, Y is continuous.

(b) The density curve will be a rectangle with base covering the region 0 to 2. The base has length 2. The area of a rectangle is base × height, so the area of the density curve is 2 × height. This area must equal 1, so the height must be 0.5.

A graph of the density curve follows.

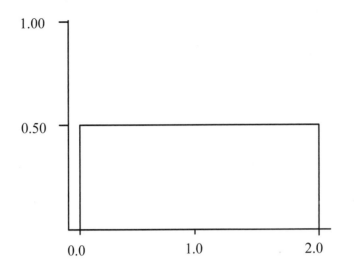

(c) Here is a graph of the desired region.

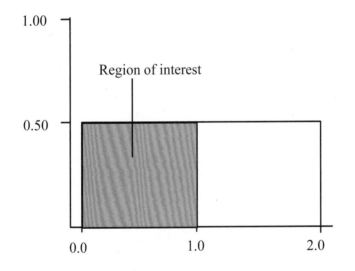

The region of interest is a rectangle with base of length 1.0 and height of 0.5. Thus, the area of this rectangle is $1 \times 0.5 = 0.5$, so

$$P(Y \leq 1) = 0.5$$

Exercise 10.51

(a) Following the methods of Chapter 3, and referring to the picture provided,

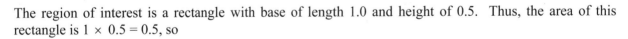

$$P(.52 \leq V \leq .60) = P(\frac{.52 - .56}{.019} \leq Z \leq \frac{.60 - .56}{.019}) = P(-2.11 \leq Z \leq 2.11) = .9826 - .0174 = .9652.$$

Note: What this says is, if 56% of all registered voters actually do vote, then when we randomly select 663 registered voters, there is a 96.5% chance that the proportion in our sample that actually voted will be between 52% and 60%.

(b) Following the methods of Chapter 3, $P(V \geq .72) = P(Z \geq \frac{.72 - .56}{.019}) = P(Z \geq 8.42)$. In other words, a sample proportion of .72 is 8.42 standard deviations away from the mean of .52. It is virtually impossible for this to occur by chance alone, so $P(V \geq .72) = 0$ (to at least four decimal places, according to the standard normal table.

So, suppose we observed such a sample proportion of .72. Then one of two things has occurred: (1) We just happened to observe something spectacularly unlikely; or (2) The percentage of all registered voters that voted is actually more than 56%, and that is why we had so many in our sample. Here, it seems that the (2) is the more plausible explanation. That is, if we observed a sample proportion of .72, we would be tempted to conclude that the proportion of all registered voters that voted is, in fact, more than .56.

Exercise 10.54

We spun a nickel 50 times and got 22 heads. From this result, we estimate the probability of heads to be

Estimate = 22/50 = 0.44

Your results will probably differ somewhat, but your estimate of the probability of heads will be the number of heads you got in 50 spins divided by 50.

Exercise 10.56

(a) Here is our sequence of hits (H) and misses (M).

H	H	M	H	H	H	M	M	H	H	H	M	H	M	H
M	H	M	M	H	M	M	H	H	H	M	M	H	H	H
M	M	M	M	H	M	M	H	H	H	H	M	H	H	H
M	M	M	M	M	H	M	H	H	M	H	M	H	M	M
H	H	H	H	M	H	M	M	M	M	H	H	M	H	H
M	M	H	H	H	M	M	H	H	M	M	H	M	H	M
M	H	M	H	H	M	H	H	H	H					

Proportion of hits = .54

(b) Go through your sequence to determine the longest string of hits or misses. In our example,

Longest run of shots hit = 4 (occurred more than once)

Longest run of shots missed = 5

CHAPTER 11

SAMPLING DISTRIBUTIONS

OVERVIEW

Statistical inference is the technique that allows us to use the information in a sample to draw conclusions about the population. Associated with any sample statistic is its **sampling distribution,** the distribution of values taken by the statistic in all possible samples of the same size from the same population. The sampling distribution can be described in the same way as the distributions you encountered in Chapters 1 and 2 of your textbook, and its three important features follow:

- Measure of center

- Measure of spread

- Description of the shape of the distribution

The statistic discussed first is the **sample mean**, $\bar{x}$, which is an estimate of the **population mean** μ. $\bar{x}$ has a number of convenient properties when taken from an SRS that allow one to make a variety of inferences about μ. As with all sampling distributions, we need to know the mean, standard deviation, and the shape of the distribution. In fact, we know from the **central limit theorem** that the shape of the distribution of the sample mean is very close to normal when simple random sampling is used and sample sizes are large. The **law of large numbers** also tells us more about the behavior of $\bar{x}$ as the sample size increases. The law of large numbers describes how the mean of many observations of a random process gets closer and closer to the mean of the population.

Here are the basic facts about the sample mean from an SRS of size n taken from a population where the mean is μ and the standard deviation is σ:

- $\bar{x}$ is an **unbiased estimate** of the population mean μ, so the mean of $\bar{x}$ is the population mean μ.

- The standard deviation is $\sigma/\sqrt{n}$, where σ is the population standard deviation. (There is less variation in averages than in individuals, and averages based on larger samples are less variable than those based on smaller samples.)

- The shape of the distribution of the sample mean depends on the shape of the population distribution. If the population was normal, $N(\mu, \sigma)$, then the sample mean has normal distribution $N(\mu, \sigma/\sqrt{n})$. Also, for a large sample the central limit theorem tells us that the sample mean has *approximately* a normal distribution $N(\mu, \sigma/\sqrt{n})$.

A statistic is called **unbiased** if the sampling distribution of the statistic is centered (has its mean) at the value of the population parameter. This means that the statistic tends to neither overestimate nor underestimate the parameter.

Statistical process control consists of methods for monitoring a process over time so that any changes in the process can be detected and corrected quickly. This is an economical method for maintaining product quality in a manufacturing process. We say that a process that continues over time is in **control** if it is operating under stable conditions. More precisely, the process is in control with respect to some variable measured on the process if the distribution of this variable remains constant over time.

Control charts are used to monitor the values of a variable measured on a process over time. One of the most common control charts is the $\bar{x}$ **control chart.** This chart is produced by periodically observing a sample of n values of the variable of interest and plotting the means $\bar{x}$ of these values versus the time order of the samples on a graph. A solid **centerline** at the target value of the process mean μ for the variable is drawn on the graph, as are dashed **control limits** at

$$\mu \pm 3\frac{\sigma}{\sqrt{n}}$$

where σ is the process standard deviation of the variable. This chart helps us decide whether the process is in control with mean μ and standard deviation σ. The probability that the next point (value of $\bar{x}$) on such a chart lies outside the control limits is about 0.003 if the process is in control. Such a point would be evidence that the process is **out of control,** that is, that the distribution of the process has changed for some reason. When a process is deemed out of control, a cause for the change in the process should be sought.

GUIDED SOLUTIONS

Exercise 11.2

KEY CONCEPTS: Statistics and parameters

In deciding whether a number represents a parameter or a statistic, you need to think about whether it is a number that describes a population of interest or whether it is a number computed from the particular sample that was selected. In this problem, what is the population? What is the sample? Based on the answer, indicate whether the number is a parameter or a statistic.

41%:

37%:

33%:

Exercise 11.6

KEY CONCEPTS: Population distribution and sampling distribution

Take care to distinguish between the population distribution (the collection of all values of a variable of interest among all individuals in the population) and the sampling distribution of a statistic (the collection of all values of the statistic in all possible samples of the same size from the same population).

What individuals are we interested in? What is the variable of interest among the individuals in this population?

(a) Describe the population and the population distribution:

Population:

Values in the population distribution:

What is the population distribution?

In this exercise, we repeatedly sample 100 individuals, each time observing and recording $\bar{x}$, the sample mean. What does each $\bar{x}$ represent?

(b) What does $\bar{x}$ represent? What do we mean by "sampling distribution of $\bar{x}$"?

Exercise 11.10

KEY CONCEPTS: Standard deviation of the sampling distribution of $\bar{x}$

(a) If $\bar{x}$ is the mean of an SRS of size n drawn from a large population with mean μ and standard deviation σ, then the standard deviation of the sampling distribution of $\bar{x}$ is $\dfrac{\sigma}{\sqrt{n}}$. To apply this formula here, identify σ and n and then complete the following:

$$\text{Standard deviation of Juan's mean result} = \frac{\sigma}{\sqrt{n}} =$$

(b) What value would n have to be so that $\dfrac{\sigma}{\sqrt{n}}$ is 5?

In this case, you know the value of σ. Hence, solve the equation $\dfrac{\sigma}{\sqrt{n}} = 5$ for n. You can do this algebraically, or you might determine the answer intuitively.

Without using formulae or technical language, write an explanation of the advantage of reporting the average of several measurements rather than the result of a single measurement. You might mention that the standard deviation of $\bar{x}$ is a measure of $\bar{x}$'s precision in estimating the true value of the thing Juan is measuring.

Exercise 11.13

KEY CONCEPTS: Central limit theorem; the four-step process

The four-step process follows:

> *State.* What is the practical question in the context of the real-world setting?
> *Plan.* What specific statistical operations does this problem call for?
> *Solve.* Make the graphs and carry out the calculations needed for this problem.
> *Conclude.* Give your practical conclusion in the setting of the real-world problem.

To apply the steps to this problem, here are some suggestions:

State. What assumption about the average loss are you asked to investigate?

Plan. You will need to assume that the 10,000 policies sold belong to homeowners who can be safely assumed to be an SRS from the population of all homeowners. What probability about the average loss ($\bar{x}$) do you need to compute?

Solve. What does the central limit theorem say about the sampling distribution of the average loss, $\bar{x}$, for 10,000 policies? Fill in the blanks.

The sampling distribution of $\bar{x}$ is approximately $N($, $)$

Now use the Normal probability calculations you learned in Chapter 3 to compute the probability that the mean $\bar{x}$ is no greater than $275.

Conclude. What does the probability you calculated in the *Solve* step tell you about whether the company can safely base its rates on the assumption that the average loss will be no greater than $275? In your opinion, is the probability sufficiently large?

Exercise 11.24

KEY CONCEPTS: Law of large numbers

Review the statement of the law of large numbers in the text. The mean payoff on a $1.00 bet is $0.947. What are the mean winnings (μ) for the gambler in this problem? The amount a gambler makes per bet on average after many bets on red would be $\bar{x}$. What does the law of large numbers say about $\bar{x}$?

Exercise 11.27

KEY CONCEPTS: Sampling distribution of a sample mean; variability of one measurement compared with variability of the mean of several measurements

Let x denote one single measure of Shelia's glucose level.
Let $\bar{x}$ denote the average of 4 measurements of Shelia's glucose level.

We know that x has mean $\mu = 125$ mg/dl and standard deviation $\sigma = 10$ mg/dl.

(a) The event that a single glucose measurement results in Shelia being diagnosed as having gestational diabetes is $\{ x > 140$ mg/dl$\}$. Hence, we need to compute $P(x > 140)$.

$P(x > 140) =$

(b) The event that the average of four separate measurements results in a diagnosis of gestational diabetes is $\{\bar{x} > 140$ mg/dl$\}$. Hence, we need to compute $P(\bar{x} > 140)$.

What is the sampling distribution of $\bar{x}$?

$P(\bar{x} > 140) =$

Exercise 11.29

KEY CONCEPTS: Sampling distribution of a sample mean; computing a value of the sample mean given a proportion

To solve Exercise 11.27 above, you needed to know the sampling distribution of $\bar{x}$, the mean of four measurements of Shelia's glucose level. You need this again to solve this problem.

What is the sampling distribution of $\bar{x}$?

Now, we need a value L such that $\bar{x}$ falls above L only 5% of the time. In other words, we need L such that $P(\bar{x} > L) = .05$. This is a problem very much like those described in the "Finding a Value Given a Proportion" section of Chapter 3. Review this section, if you need to.

Notice that if $\bar{x}$ is larger than L only 5% of the time, then it's less than L 95% of the time.

What z-score does L have?

The value of L is this many standard deviations above the average. Use this to compute L:

$$L =$$

Exercise 11.38

KEY CONCEPTS: Central limit theorem; variability of the sampling distribution of the sample mean when the population distribution is highly variable

Let x denote Joe's payoff on a single $1 wager. We know that x has mean $\mu = \$0.60$ and standard deviation $\sigma = \$18.96$. Over 40 years, Joe plays $n = 14{,}000$ times.

Let $\bar{x}$ denote the mean payoff over Joe's long career of 14,000 wagers.

(a) Use the formulas for mean and standard deviation of a sample mean to compute the following:

Mean of $\bar{x}$:

Standard deviation of $\bar{x}$:

(b) According to the central limit theorem, the approximate sampling distribution of $\bar{x}$ follows the Normal distribution with mean and standard deviation computed in (a). Use this to compute $P(.50 < \bar{x} < .70)$.

$$P(.50 < \bar{x} < .70) =$$

Exercise 11.40

KEY CONCEPTS: Accuracy of central limit theorem approximation to sampling distribution of the sample mean.

Let $\bar{x}$ be Joe's average payoff based on n wagers. In this problem, we consider three cases for the value of n: 14,000, 3,500, and 150,000. In all three cases, we will use the central limit theorem to approximate $P(.50 < \bar{x} < .70)$. In all three cases, we are also given the exact value of this probability, obtained using methods outside the scope of this text. We'll compare the exact probability to the central limit theorem based approximation and see how the accuracy of this approximation depends on the sample size, n.

(a) $n = 14,000$. This is the scenario described in Exercise 11.38.

What is the approximate sampling distribution of $\bar{x}$?

Use this to approximate $P(.50 < \bar{x} < .70)$. Your answer from part (b) of Exercise 11.38:

The exact probability: $P(.50 < \bar{x} < .70) = 0.4961$

How accurate is the normal approximation? You might compute the absolute value of the difference between the exact probability and the central limit theorem approximation.

(b) $n = 3,500$

What is the approximate sampling distribution of $\bar{x}$? The standard deviation of $\bar{x}$ should change.

Use this to approximate $P(.50 < \bar{x} < .70)$:

The exact probability: $P(.50 < \bar{x} < .70) = 0.4048$

How accurate is the Normal approximation?

(c) $n = 150,000$

What is the approximate sampling distribution of $\bar{x}$?

Use this to approximate $P(.50 < \bar{x} < .70)$:

The exact probability: $P(.50 < \bar{x} < .70) = 0.9629$

How accurate is the normal approximation?

COMPLETE SOLUTIONS

Exercise 11.2

In this problem, the population consists of all registered voters in Florida. Of these, 41% are registered as Democrats and 37% as Republicans. Since these numbers describe all of the voters, they're both parameters.

The sample described in this problem consists of the registered voters among 250 people contacted by randomly selecting 250 residential telephone numbers. Of these, 33 are registered Democrats. This number describes the sample, not the population of all voters… so it's a statistic.

Exercise 11.6

(a) The *population* consists of the 12,000 able-bodied undergraduate males enrolled at the University of Illinois that participated in this study. The *population distribution* consists of the 1-mile run times of the individuals in this population. As described, this population distribution follows the Normal distribution wit mean 7.11 minutes and standard deviation 0.74 minutes.

(b) We randomly sample 100 of these run times (by sampling 100 men) and the compute their average run time ($\bar{x}$). In this question, we envision repeatedly observing $\bar{x}$'s, each based on 100 randomly selected run times. The sampling distribution of $\bar{x}$ then consists of all values of $\bar{x}$ taken by all possible samples of 100 randomly selected run times.

Exercise 11.10

(a) In this exercise $\sigma = 10$ and $n = 3$. Thus, the standard deviation of Juan's mean result is
$$\frac{\sigma}{\sqrt{n}} = \frac{10}{\sqrt{3}} = 5.77.$$

(b) We want $\dfrac{\sigma}{\sqrt{n}} = \dfrac{10}{\sqrt{n}}$ to be equal to 5. Hence, we solve $\dfrac{10}{\sqrt{n}} = 5$ for n.

Algebraically: Solving this equation, we have $\sqrt{n} = 10/5 = 2$. Squaring both sides yields $n = 4$.

Intuitively: It's clear that when $n = 4$, $\dfrac{10}{\sqrt{n}} = \dfrac{10}{\sqrt{4}} = \dfrac{10}{2} = 5$, as desired.

Averages of several measurements are more likely than a single measurement to be closer to the true value of the quantity being measured. This is because more measurements reduce the magnitude of chance deviations or errors of estimation in the sample mean, $\bar{x}$. In other words, $\bar{x}$ is more likely to be close to the true value being measured than a single measurement is.

Exercise 11.13

State. We need to decide whether the company can safely sell 10,000 such policies on the assumption that the average losses on those policies will be no greater than $275.

Plan. We need to calculate the probability that the average annual loss, $\bar{x}$, for these 10,000 policies will be no greater than $275. If this probability is very large, then it's safe to assume that $\bar{x}$ will be no more than $275. {You could also compute the chance that $\bar{x}$ is *more* than $275. If this probability is small enough, then we would reach the same conclusion.}

Solve. We are told that the population of all homeowners has a mean annual loss of $250 and the standard deviation of the loss is $1000. The central limit theorem tells us that the sampling distribution of $\bar{x}$ is approximately Normal with mean $250 (the same as the population mean) and standard deviation

$$\frac{\sigma}{\sqrt{n}} = \frac{\$1000}{\sqrt{10,000}} = \frac{\$1000}{100} = \$10$$

Thus, the sampling distribution of $\bar{x}$ is approximately $N(\$250, \$10)$.

The probability that $\bar{x}$ is no greater than $275 is

$$P(\bar{x} \le \$275) = P(\frac{\bar{x} - \$250}{\$10} \le \frac{\$275 - \$250}{\$10}) = P(Z \le 2.5) = 0.9938$$

according to our standard Normal table.

Conclude. The probability that its average loss will be no greater than $275 is 0.9938. {By Probability Rule 4 (Chapter 10), the probability that its average loss will be *more* than $275 is therefore $1 - .9938 = .0062$.} As with all statistical methods, the final decision as to whether this number (0.9938) is close enough to 1 for the assumption in question to be considered safe is a subjective one. However, perhaps most would conclude that the assumption may be considered safe. The issue of reaching conclusions based on probabilities such as this will be one of the major concepts discussed in Chapter 14.

Exercise 11.24

The gambler pays $1.00 for an expected payout of $0.947. His mean *winnings* are therefore $0.947 − $1.00 = −$0.053 per bet. In other words, the expected *losses* of the gambler are $0.053 per bet. The law of large numbers tells us that if the gambler makes a large number of bets on red, keeps track of his net winnings, and computes the average $\bar{x}$ of these, this average will be close to −$0.053. In other words, he will find, in the long run, that he loses about 5.3 cents per bet on average. The law of large numbers refers to the "long run average" - in the long run, our gambler loses 5.3 cents per spin of the wheel when placing such a bet. But, in any fixed number of wagers, the average loss ($\bar{x}$) for those bets will differ from 5.3 cents. Indeed, it's possible for the player to come out ahead, but only in the short run. As the number of wagers increases, it becomes ever less likely for the average loss ($\bar{x}$) to stray from 5.3 cents per wager.

Exercise 11.27

(a) Let x denote one single measure of Shelia's glucose level. We know that x has the Normal distribution with mean $\mu = 125$ mg/dl and standard deviation $\sigma = 10$ mg/dl. Hence, using the methods for computing Normal probabilities described in Chapter 3,

$$P(x > 140) = P(\frac{x - 125}{10} > \frac{140 - 125}{10}) = P(Z > 1.5) = 1 - .9332 = .0668$$

If Shelia's average measured glucose level is really 125 mg/dl, the chance that she gets labeled as having gestational diabetes due to one measurement over 140 mg/dl is about 6.68%.

(b) Let $\bar{x}$ be the average of four measurements. Then $\bar{x}$ has a Normal distribution with mean $\mu = 125$ mg/dl and standard deviation $\dfrac{\sigma}{\sqrt{n}} = \dfrac{10}{\sqrt{4}} = 5$ mg/dl.

$$P(\bar{x} > 140 \text{ mg/dl}) = P(\frac{\bar{x} - 125}{5} > \frac{140 - 125}{5}) = P(Z > 3) = 1 - .9984 = .0016$$

This means that if Shelia's actual average measured glucose level one hour after a sugary drink is 125 mg/dl, it is highly unlikely that the average of four measurements would classify her as a gestational diabetic.

Exercise 11.29

In Exercise 11.27, we found that the sampling distribution of the mean ($\bar{x}$) of four glucose measurements Normal with mean $\mu = 125$ mg/dl and standard deviation 5 mg/dl.

We seek L such that $P(\overline{X} > L) = .05$.

Referring to the table of standard Normal probabilities, and using the methods described in Chapter 3, we want the value of Z with .9500 area under the curve to the left. There are two values in the table equally close to .9500: $Z = 1.64$ yields area .9495 to the left, while $Z = 1.65$ yields area .9505 to the left. We could use either $Z = 1.64$ or $Z = 1.65$, or we could average them and use $Z = 1.645$. For simplicity, let's say that $Z = 1.65$. Hence, L has a z-score of 1.65.

So L must be 1.65 standard deviations above average: $L = 125$ mg/dl $+ 1.65 \times 5$ mg/dl $= 133.25$ mg/dl. If you use $Z = 1.64$ or $Z = 1.645$, the answer will change little.

Exercise 11.38

(a) Using the formulas for mean and standard deviation of a sample mean to compute the following:

$$\text{Mean of } \bar{x} = \$0.60$$

$$\text{Standard deviation of } \bar{x} = \frac{\sigma}{\sqrt{n}} = \frac{\$18.96}{\sqrt{14,000}} = \$0.16024$$

(b) According to the central limit theorem, the approximate sampling distribution of $\bar{x}$ follows the Normal distribution with mean $0.60 and standard deviation $0.16024. Hence,

$$P(.50 < \bar{x} < .70) = P\left(\frac{.50-.60}{.16024} < \frac{\bar{x}-0.60}{.16024} < \frac{.70-.60}{.16024}\right) = P(-.62 < Z < .62) = 0.7324 - 0.2676 = 0.4648$$

It might be surprising to realize that even after 14,000 plays of this wager, Joe has a better than 50% chance to observe an average payoff more than 10 cents from the long run mean payoff.

Exercise 11.40

(a) As shown in Exercise 11.38, when $n = 14,000$:
Central limit theorem approximation: $P(.50 < \bar{x} < .70) = .4648$
Exact probability: $P(.50 < \bar{x} < .70) = .4961$
Size of error of approximation $= |.4648 - .4961| = .0313$

(b) If $n = 3,500$, then the standard deviation of $\bar{x}$ is $\frac{\sigma}{\sqrt{n}} = \frac{\$18.96}{\sqrt{3500}} = \$0.32048$. Notice (comparing (b) to (a)), when you quadruple the sample size, the standard deviation of the sample mean is halved.
Central limit theorem approximation:
$$P(.50 < \bar{x} < .70) = P\left(\frac{.50-.60}{.32048} < \frac{\bar{x}-0.60}{.32048} < \frac{.70-.60}{.32048}\right) = P(-.31 < Z < .31) = .6217 - .3783 = .2434$$
Exact probability: $P(.50 < \bar{x} < .70) = .4048$
Size of error of approximation $= |.4048 - .2434| = .1614$
This error is so large, you might wonder if a mistake has been made. The population so variable and highly skewed (lots of 0's, a few $600 payoffs), it takes a very, very large sample for the sample mean to have a sampling distribution that is well approximated by the Normal distribution. This explains why the Normal approximation here is so poor.

(c) If $n = 150,000$, then the standard deviation of $\bar{x}$ is $\frac{\sigma}{\sqrt{n}} = \frac{\$18.96}{\sqrt{150,000}} = \$0.04895$.

Central limit theorem approximation:
$$P(.50 < \bar{x} < .70) = P\left(\frac{.50-.60}{.04895} < \frac{\bar{x}-0.60}{.04895} < \frac{.70-.60}{.04895}\right) = P(-2.04 < Z < 2.04) = .9793 - .0207 = .9586$$
Exact probability: $P(.50 < \bar{x} < .70) = .9629$
Size of error of approximation $= |.9629 - .9586| = .0043$
With a sample so large, the Normal approximation provided by the central limit theorem is very good.

CHAPTER 12

GENERAL RULES OF PROBABILITY

OVERVIEW

Chapter 6 of your text discusses two-way tables and conditional distributions. Here, we learn about **conditional probabilities** and their use in calculating probabilities of complex events. The conditional probability of an event B given that an event A has occurred is denoted $P(B \mid A)$ and is defined by

$$P(B \mid A) = \frac{P(A \text{ and } B)}{P(A)}$$

where $P(A) > 0$. In practice, a conditional probability can often be determined directly from the information given in a problem. Events are **independent** if knowledge that one event has occurred does not alter the probability that the second event occurs. Specifically, two events A and B are independent if $P(B \mid A) = P(B)$. It follows that for independent events we must have the result that $P(A \text{ and } B) = P(A)P(B)$. In any particular problem, we can use this result to check if two events are independent by determining if the probabilities multiply correctly. However, independence is usually assumed as part of the probability model.

The following general rules are valid for any assignment of probabilities and allow us to compute the probabilities of events in many random phenomena.

Addition rule: If events A, B, and C are all disjoint in pairs, then

$$P(\text{at least one of these events occur}) = P(A) + P(B) + P(C)$$

Multiplication rule: If events A, B, and C are independent, then

$$P(\text{all of the events occur}) = P(A)P(B)P(C)$$

General addition rule: For any two events A and B,

$$P(A \text{ or } B) = P(A) + P(B) - P(A \text{ and } B)$$

General multiplication rule: For any two events A and B,

$$P(A \text{ and } B) = P(A)P(B \mid A)$$

GUIDED SOLUTIONS

Exercise 12.1

KEY CONCEPTS: Independence, multiplication rule

The key concept that must be properly understood to answer the questions raised in this exercise is the notion of independence. Let A be the event an adult is a full-time college student and B be the event that an adult is 55 years or older. Events A and B are independent if knowledge that A has occurred (full time college student) does not alter our assessment of the probability that B will occur (person is over 55 years old).

What assumption has been made to use the multiplication rule to find the probability that A and B occur? Do you think the multiplication rule applies here?

Exercise 12.2

KEY CONCEPTS: Independence, multiplication rule

What is the probability that a single textbook would have an author whose name was not among the ten most common names?

What is the probability that all nine textbooks would have an author whose name was not among the ten most common names?

Exercise 12.5

KEY CONCEPTS: Venn diagrams

(a) Let A be the event that the student is 25 years old or older and B be the event that the student is local. Make a Venn diagram in the space below similar to the Venn diagram in Figure 12.4, which shows the four events.

(b) Describe each of the events in words. We have written out one of the descriptions as an example.

{*A* and *B*}

{*A* and not *B*} – the randomly selected student is older than 25 and not local

{*B* and not *A*}

{neither *A* nor *B*}

(c) Find the probabilities of these four events and add them to the Venn diagram that you have drawn in part (a).

Exercise 12.7

KEY CONCEPTS: Conditional probability

Let *A* be the event that the student is 25 years old or older and *B* be the event that the student is local. You are asked to find the conditional probability that a student is local given that he or she is less than 25 years old, namely, $P(B \mid \text{not } A)$. Applying the general formula for conditional probabilities,

$$P(B \mid \text{not } A) = \frac{P(B \text{ and not } A)}{P(\text{not } A)} =$$

Use the information from Exercise 12.5 to first evaluate the probabilities $P(B \text{ and not } A)$ and $P(\text{not } A)$. Then, use these probabilities to compute the conditional probability above.

Exercise 12.30

KEY CONCEPTS: Multiplication rule for independent events

(a) The three years are independent. If *U* indicates a year for the price being up and *D* indicates a year for the price being down, you need to compute $P(UUU)$.

(b) This problem must be set up carefully and done in steps.

Step 1. Write the event of interest in terms of simpler outcomes. How would you write $P(UUU \text{ or } DDD)$ in terms of $P(UUU)$ and $P(DDD)$?

$P(\text{moves in the same direction in the next two years}) = P(UUU \text{ or } DDD) =$

Step 2. Evaluate $P(UUU)$ and $P(DDD)$ and substitute your answer in the expression from Step 1.

Exercise 12.37

KEY CONCEPTS: Conditional probabilities, geometric probabilities

You want to use geometric arguments to evaluate

$$P(Y < 1/2 \mid Y > X) = \frac{P(Y < 1/2 \text{ and } Y > X)}{P(Y > X)}$$

In the previous equation, we have just used the definition of conditional probability where A corresponds to "$Y < 1/2$" and B corresponds to "$Y > X$." The three figures that follow are of the square $0 \le x \le 1$ and $0 \le y \le 1$. In the figure on the left, the region corresponding to "$Y > X$" is shaded. Remembering that probabilities correspond to areas, first evaluate $P(Y > X)$.

$P(Y > X) =$

In the middle figure the region corresponding to "$Y < 1/2$" is shaded, and in the figure on the right the region "$Y < 1/2$ and $Y > X$" is shaded. The area of the shaded triangle in the figure on the right corresponds to the probability $P(Y < 1/2 \text{ and } Y > X)$. The value of this probability is

$P(Y < 1/2 \text{ and } Y > X) =$

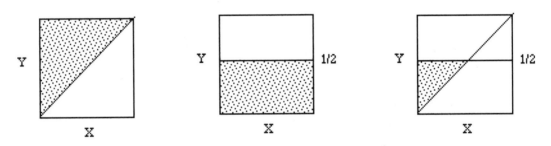

Putting this all together, you should be able to evaluate the following probability.

$$P(Y < 1/2 \mid Y > X) = \frac{P(Y < 1/2 \text{ and } Y > X)}{P(Y > X)} =$$

Exercise 12.39

KEY CONCEPTS: Two-way table of counts, conditional probabilities

Following is the table from the exercise.

	Bachelor's	Master's	Professional	Doctorate	Total
Female	986	411	52	32	1481
Male	693	260	45	27	1025
Total	1679	671	97	59	2506

(a) You can calculate this probability directly from the table. All degree recipients in the table are equally likely to be chosen (that is what it means to choose a degree recipient at random), so the fraction of the degree recipients in the table who are women is the desired probability. How many women degree recipients are there? Where do you find this number in the table? What is the total number of degree recipients represented in the table? Use these numbers to compute the desired fraction.

P(choose a woman) =

(b) This probability can also be calculated directly from the table. Since this is a conditional probability (this is a probability given that the degree recipient receives a doctorate), we restrict ourselves only to doctorate degree recipients. The desired probability is then the fraction of these doctorate degree recipients who are women. Find the appropriate entries in the table to compute this fraction.

P(choose a woman | choose a doctorate degree recipient) =

(c) If the two events "choose a woman" and "choose a person receiving a doctorate" are independent, then what should be true about the probabilities computed in (a) and (b)? Are these events independent?

Exercise 12.49

KEY CONCEPTS: Multiplication rules and conditional probability, tree diagrams

We are given the following probabilities:

P(white) = 0.40

P(black) = 0.40

P(hispanic) = 0.20

P(vote for candidate | white) = 0.30

P(vote for candidate | black) = 0.90

P(vote for candidate | hispanic) = 0.50

Draw a tree diagram as in Figure 12.5 to organize the information and to compute P(vote for candidate). If you are having difficulty, this exercise is very similar to Example 12.10 in your text.

120   Chapter 12

Exercise 12.51

KEY CONCEPTS: Conditional probability

We want to calculate the conditional probability $P(\text{black} \mid \text{vote for candidate})$. Using the general formula for conditional probability, we have

$$P(\text{black} \mid \text{vote for candidate}) = \frac{P(\text{black and vote for candidate})}{P(\text{vote for candidate})}$$

In Exercise 12.49, you computed $P(\text{vote for candidate})$. Use the tree diagram in Exercise 12.49 or the general multiplication rule to evaluate the numerator:

    $P(\text{black and vote for candidate}) =$

You can now use these two probabilities to compute the desired conditional probability:

    $P(\text{black} \mid \text{vote for candidate}) =$

Exercise 12.55

KEY CONCEPTS: Independence, multiplication rule

(a) The possible alleles inherited are B and B, B and O, and O and O. What blood types do these inherited alleles result in?

(b) Let S_O and S_B correspond to the events that allele O or B is inherited from Sarah, respectively, and D_O and D_B correspond to the events that allele O or B is inherited from David. S_O and S_B each has probability 0.5; D_O and D_B each has probability 0.5.

$P(\text{child has type O}) = P(S_O \text{ and } D_O) =$

What rule allows you to multiply the probabilities?

How many blood types can their children have? What is the probability that the child has type B blood?

COMPLETE SOLUTIONS

Exercise 12.1

Suppose the events "full-time student" and "over 55 years of age" are independent. This would imply that knowing whether an adult was a full time student would not change the probability that the adult was over 55 years old. In terms of this problem, if the events were independent, 8% of adults would be full-time students and 8% of the adults over 55 would be full-time college students (knowledge of whether or not an adult was over 55 doesn't alter – increase or decrease – their chance of being a full time student). The independence is what allows us to just multiply these probabilities together. The use of the formula $(0.08)(0.30) = 0.024$ to get the answer requires that 8% of adults over 55 are full-time students. However, we would guess that fewer than 8 % of those over 55 are full-time college students, so that multiplying the two probabilities together is not the correct way to get the answer. The answer obtained by multiplying the two probabilities together is too large. –

Exercise 12.2

The probability that a single textbook would have an author whose name was not among the ten most common names is $1 - 0.096 = 0.904$, assuming that a person's name is not related to whether or not they write a textbook.

The probability that all nine textbooks would have an author whose name was not among the ten most common names has probability

P(all 9 textbooks have authors whose name is not among the 10 most common) $= (0.904)^9 = 0.4032$

This probability assumes that the names of the authors of different textbooks are independent of each other. Thus, it would not be that surprising if none of the names of these authors were among the ten most common.

Exercise 12.5

(a) The four events are shown in the Venn diagram below. The larger rectangle in the lower left corresponds to the event A, and the smaller rectangle corresponds to the event B.

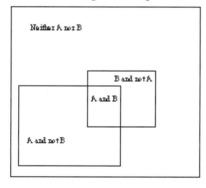

(b) The events described in words are
{A and B} – the randomly selected student is older than 25 and local
{A and not B} – the randomly selected student is older than 25 and not local
{B and not A} – the randomly selected student is local and younger than 25
{neither A nor B} – the randomly selected student is younger than 25 and not local

(c) We are told that $P(A) = 0.7$, $P(B) = 0.25$, and $P(A \text{ and } B) = 0.05$. Since 70% of students are older than 25 and 5% are older than 25 and local, we must have $0.7 - 0.05 = 0.65$ in the region corresponding to older than 25 and not local, that is, $\{A \text{ and not } B\}$. Since 25% of students are local and 5% are local and older than 25, we must have $0.25 - 0.05 = 0.2$ in the region corresponding to local and younger than 25, that is $\{B \text{ and not } A\}$. Finally, the region outside the shapes corresponding to older than 25 and local, that is, $\{$neither A nor $B\}$, must have 10% of the students since the four regions in the Venn diagram are disjoint and make up the entire sample space.

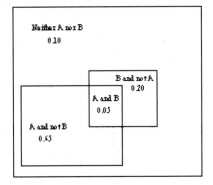

Exercise 12.7

From Exercise 12.5 we know that $P(B \text{ and not } A) = 0.2$ and $P(\text{not } A) = 1 - P(A) = 1 - 0.7 = 0.3$. Plugging these values into the formula for conditional probability yields

$$P(B \mid \text{not } A) = \frac{P(B \text{ and not } A)}{P(\text{not } A)} = \frac{0.2}{0.3} = 0.667$$

Exercise 12.30

(a) $P(UUU) = (0.65)^3 = 0.2746$

(b) Because the events UUU and DDD are disjoint

$P(\text{moves in the same direction in the next two years}) = P(UUU \text{ or } DDD) = P(UUU) + P(DDD)$.

From part (a), $P(UUU) = 0.2746$. The probability of the price being down in any given year is $1 - 0.65 = 0.35$. Since the years are independent, the probability of the price being down in three consecutive years is $P(DDD) = (0.35)^3 = 0.0429$. Putting this together,

$P(\text{moves in the same direction in the next two years}) = 0.2746 + 0.0429 = 0.3175$

Exercise 12.37

The figure on the left in the Guided Solution has the region corresponding to "$Y > X$" shaded, and because probabilities correspond to areas, we see that $P(Y > X) = 1/2$. In the figure on the right, the region "$Y < 1/2$ and $Y > X$" is shaded. The area of the shaded triangle in the figure on the right is $1/8$, so $P(Y < 1/2 \text{ and } Y > X) = 1/8$.

Putting this all together gives

$$P(Y<1/2|Y>X) = \frac{P(Y<1/2 \text{ and } Y>X)}{P(Y>X)} = \frac{1/8}{1/2} = 0.25$$

Exercise 12.39

(a) The number of women degree recipients is found as the total for the first row and is (in thousands) 1481. The total number of degree recipients in the table is in the lower right corner and is (in thousands) 2506. The desired probability is thus

P(choose a woman) = (number of women degree recipients) / (total number of recipients in table)

$$= 1481/2506 = 0.5910$$

(b) The desired conditional probability is

P(choose a woman | choose a doctorate degree recipient)

= (number of doctorate degree recipients who are women) / (number of doctorate degree recipients)

= 32/59 = 0.5424

(c) If the two events "choose a woman" and "choose a doctorate degree recipient" are independent, then we should have

$$P(\text{choose a woman}) = P(\text{choose a woman} \mid \text{choose a doctorate degree recipient})$$

These are the two probabilities that you computed in (a) and (b). Since they are not equal, these two events are not independent.

Exercise 12.49

The tree diagram below organizes the information given in the problem.

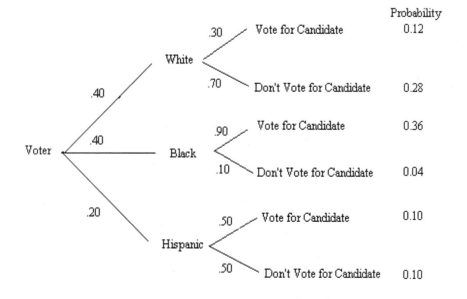

A voter is either white, black, or Hispanic. The proportion of voters of each race mark the three leftmost branches in the tree. Look at the top branch corresponding to the white. The two segments going out from the "white" branch point have the conditional probabilities:

$$P(\text{vote for candidate} \mid \text{white}) = 0.30$$

$$P(\text{don't vote for candidate} \mid \text{white}) = 0.70$$

Now use the multiplication rule to find the probability that a white voter votes for the candidate:

$$P(\text{white and votes for candidate}) = P(\text{white}) \, P(\text{votes for candidate} \mid \text{white})$$

$$= (0.40)(0.30) = 0.12$$

This probability appears at the end of the topmost branch. The probabilities of all six complete branches are computed in this manner. There are three paths leading to "vote for candidate", and these paths are disjoint. Thus, the percentage of the overall vote that the candidate gets is the sum of the probabilities associated with these three disjoint paths:

$$P(\text{vote for candidate}) = 0.12 + 0.36 + 0.10 = 0.58.$$

Exercise 12.51

The conditional probability of interest is

$$P(\text{black} \mid \text{vote for candidate}) = \frac{P(\text{black and vote for candidate})}{P(\text{vote for candidate})}$$

From the tree diagram, the event "black and vote for candidate" is the third branch from the top of the tree and has

$$P(\text{black and vote for candidate})$$

$$= P(\text{vote for candidate} \mid \text{black})P(\text{black})$$

$$= (0.9)(0.40) = 0.36.$$

From Exercise 12.49, $P(\text{vote for candidate}) = 0.58$. Putting them together,

$$P(\text{black} \mid \text{vote for candidate}) = \frac{0.36}{0.58} = 0.621$$

Approximately 62% of the candidate's votes come from black voters.

Exercise 12.55

(a) The possible alleles inherited are B and B, B and O, and O and O. The alleles B and B and B and O both result in a blood type of B for a child. The alleles O and O result in a blood type of O for a child. So the two blood types their children can have are B and O.

(b) Let S_O and S_B correspond to the events that allele O or B is inherited from Sarah, respectively, and D_O and D_B correspond to the events that allele O or B is inherited from David. N_O and N_B each have probability 0.5, and so do D_O and D_B:

$$P(\text{child has type O}) = P(S_O \text{ and } D_O) = 0.5 \times 0.5 = 0.25$$

You multiply the probabilities because we inherit alleles independently from our mother and father. Since the child must have blood type B or O, the $P(\text{child has type B}) = 1 - P(\text{child has type O}) = 1 - 0.25 = 0.75$.

CHAPTER 13

BINOMIAL DISTRIBUTIONS

OVERVIEW

One of the most common situations giving rise to a **count** X is the **binomial setting.** The binomial setting consists of four assumptions about how the count was produced:

- The number n of observations is fixed.
- The n observations are all independent.
- Each observation falls into one of two categories called "success" and "failure."
- The probability of success p is the same for each observation.

When these assumptions are satisfied, the number of successes, X, has a **binomial distribution** with n trials and success probability p. For smaller values of n, the probabilities for X can be found easily using statistical software or the exact **binomial probability formula.** The formula is given by

$$P(X = k) = \binom{n}{k} p^k (1-p)^{n-k}$$

where $k = 0, 1, 2, \cdots, n$, and $\binom{n}{k} = \dfrac{n!}{k!(n-k)!}$ is called the **binomial coefficient.**

When the population is much larger than the sample, a count X of successes in an SRS of size n has approximately the binomial distribution with n equal to the sample size and p equal to the proportion of successes in the population.

The mean of a binomial random variable X is

$$\mu = np$$

and the standard deviation is

$$\sigma = \sqrt{np(1-p)}$$

When n is large, the count X is approximately $N\left(np, \sqrt{np(1-p)}\right)$. This approximation should work well when $np \geq 10$ and $n(1-p) \geq 10$.

GUIDED SOLUTIONS

Exercise 13.1

KEY CONCEPTS: Binomial setting

Four assumptions need to be satisfied to ensure that the count X has a binomial distribution. The number of observations or trials is fixed in advance, each trial results in one of two outcomes, the trials are independent, and the probability of success is the same from trial to trial. In addition, for a large population with a proportion p of successes, we can use the binomial distribution as an approximation to the distribution of the count X of successes in an SRS of size n. For the setting in this exercise, see if all four assumptions are satisfied.

Think about whether random-digit dialing fits in the binomial setting. What is n? What are the two outcomes, and why might the trials be considered independent?

Exercise 13.3

KEY CONCEPTS: Binomial setting

Review the binomial assumptions. Think about whether the success probability is fixed.

Exercise 13.5

KEY CONCEPTS: Binomial probabilities, binomial tables

(a) Suppose we let X denote the number of errors caught. There are 10 word errors in the essay, and we are in the binomial setting with $n = 10$ trials. Letting "success" correspond to the students catching a word error, we have $p = 0.7$. The distribution of the number of errors caught is $B(10, 0.7)$.

Suppose Y denotes the number of errors missed. What are the values of n and p? What is the distribution of the number of errors missed?

(b) Y, the number of misses, has the binomial distribution with $n = 10$ and $p = 0.3$. You need to find the probability that $Y = 3$. The exact binomial probability formula is given by

$$P(Y = k) = \binom{n}{k} p^k (1 - p)^{n-k}$$

and the required probability can be found by plugging the appropriate values of n, k, and p into the formula. Do this to evaluate

$$P(Y = 3) =$$

Alternatively, software (or many calculators) can be used to evaluate binomial probabilities such as $P(X = 3)$, although you should do at least one computation by hand to make sure you understand how to use the formula. If you have software, use it to evaluate the probability of missing 3 or more out of 10, or

$$P(Y \geq 3) =$$

Exercise 13.9

KEY CONCEPTS: Binomial distribution, mean, and variance

(a) Suppose we let X denote the number of errors missed. What is the probability of missing an error? The distribution of X is binomial – fill in the parameters below.

$$B(\quad , \quad)$$

Now suppose we let Y denote the number of errors caught. What is the probability of catching an error? The distribution of X is binomial – fill in the parameters below.

$$B(\quad , \quad)$$

(b) Suppose we let Y denote the number of errors caught. The distribution of the number of errors caught is $B(10, 0.7)$. The mean of X is

$$\mu = np =$$

Now, suppose X denotes the number of errors missed. The mean of X is

$$\mu = np =$$

What is the total of the number of errors caught plus the number of errors missed? Can you see why the means must add to 10?

(c) If Y is the number of errors caught, the standard deviation of Y is

$$\sigma = \sqrt{np(1-p)} =$$

If X is the number of errors missed, the standard deviation of X is

$$\sigma = \sqrt{np(1-p)} =$$

Can you see why the standard deviation of the count of successes and the count of failures is the same?

Exercise 13.11

KEY CONCEPTS: Mean and standard deviation of the binomial, Normal approximation for counts

(a) Let X denote the number of students that will attend among the 1535 students admitted. The problem tells us to treat X as a count having the $B(1535, 0.27)$ distribution because the probability of a student deciding to attend is 0.27 and the students decide independently. We are asked to compute the mean μ and standard deviation σ of X. What are the formulas for μ and σ? To answer this, you may want to refer to the Overview for this chapter. Use these formulas to compute the mean and standard deviation.

$\mu =$

$\sigma =$

(b) Compute $P(X \geq 416)$ using the Normal approximation to the binomial distribution. This is the approximate probability that the college gets more students than they want. First, check to see that np and $n(1-p)$ are both greater than or equal to 10.

This is a Normal probability calculation like those discussed in Chapter 3. You may want to review the material there to refresh your memory on how to do such calculations. The first step is to standardize the number 416 (compute its z-score) by subtracting the mean μ and dividing the result by σ. Next, use Table A to determine the area to the right of this z-score. It may be helpful to draw a Normal curve to visualize the area.

$P(X \geq 416) =$

(c) Compute the probability at least 416 students accept using statistical software. How good is the approximation?

Exercise 13.25

KEY CONCEPTS: Binomial probabilities, mean and standard deviation of a binomial count

(a) How many trials are there? If a success is a rise in the index, what is the success probability?

$n =$ $\qquad$ $p =$

(b) What are the possible values of X?

(c) Use either the exact binomial formula or statistical software to evaluate the probabilities of each possible value of X and then draw a probability histogram for the distribution of X below. You may want to refer to Exercise 13.5 of this Study Guide if you are having difficulties.

$P(X = 0) =$

$P(X = 1) =$

$P(X = 2) =$

$P(X = 3) =$

$P(X = 4) =$

$P(X = 5) =$

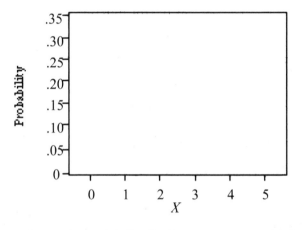

(d) When X is known to have a binomial distribution, you can use the formulas that express the mean and standard deviation of X in terms of n and p. Make sure to mark the location of the mean on your histogram.

Mean =

Standard deviation =

COMPLETE SOLUTIONS

Exercise 13.1

The number of trials, 15, is fixed and each call either succeeds in talking with a live person or it doesn't. Whether or not a particular call succeeds will not alter the probability that any other call succeeds, and the probability of any call succeeding is given as 0.20. The binomial distribution with $n = 15$ and $p = 0.20$ should be a good probability model for the number of calls that reach a live person.

Exercise 13.3

The success probability doesn't remain fixed as the student gets additional instruction if the initial answer that they give is wrong.

Exercise 13.5

(a) The distribution of the number of errors caught is given in the Guided Solutions. If Y denotes the number of errors missed, then $n = 10$ and $p = 1 - 0.7 = 0.3$, where p is now the probability of missing a given error. The distribution of the number of errors missed is $B(10, 0.3)$.

(b) Letting Y denote the number of errors missed, we want $P(Y = 3)$. Using the binomial formula to evaluate this probability, we have $n = 10$, $k = 3$, and $p = 0.3$, so that

$$P(Y = k) = \binom{n}{k} p^k (1-p)^{n-k} = \binom{10}{3} .3^3 (1-.3)^{10-3} = \frac{10!}{3!7!}(0.3)^3(0.7)^7 = 120(0.027)(0.08235) = 0.2668$$

Using software, we have $P(Y \geq 3) = 0.6172$.

Exercise 13.9

(a) If X denotes the number of errors missed, the distribution of X is $B(10, 0.3)$. If Y denotes the number of errors caught, the distribution of Y is $B(10, 0.7)$.

(b) If Y denotes the number of errors caught, the mean of Y is $\mu = np = 10(0.7) = 7$. Suppose X denotes the number of errors missed. The mean of X is $\mu = np = 10(0.3) = 3$. We see that these means add to 10. In any experiment, the total of the number of errors caught plus the number of errors missed must *always* be 10, so 10 must be the mean of this total.

(c) If Y is the number of errors caught, the standard deviation of Y is

$$\sigma = \sqrt{np(1-p)} = \sqrt{10(0.7)(0.3)} = 1.4491$$

If X is the number of errors missed, the standard deviation of X is

$$\sigma = \sqrt{np(1-p)} = \sqrt{10(0.3)(0.7)} = 1.4491,$$

So we see that the standard deviation of the count of successes and the count of failures is the same.

Exercise 13.11

(a) $\mu = np = 1535 \times 0.27 = 414.45$ and $\sigma = \sqrt{np(1-p)} = \sqrt{1535 \times 0.27 \times 0.73} = \sqrt{302.549} = 17.394$

(b) First, we check that

$$np = 1535 \times 0.27 = 414.45 \geq 10 \text{ and } n(1-p) = 1535 \times 0.83 = 1120.55 \geq 10$$

When n is large, X is approximately $N\left(np, \sqrt{np(1-p)}\right) = N(414.45, 17.394)$. Thus,

$$z\text{-score of } 416 = \frac{416 - 414.45}{17.394} = 0.089$$

and using Table A,

$$P(X \geq 416) = P(Z \geq 0.09) = 1 - P(Z \leq 0.09) = 1 - 0.5359 = 0.4641$$

(c) Using Minitab, $P(X \geq 416) = 1 - P(X \leq 415) = 1 - 0.5258 = 0.4742$. The approximation is fairly close.

Exercise 13.25

(a) X has a binomial distribution with $n = 5$ (the number of years to be observed) and $p = 0.65$ (the probability the index will increase in any given year). The independence of years is assumed as part of the model.

(b) Because $n = 5$, the possible values are X are 0, 1, 2, 3, 4, 5.

(c) To calculate the probability of each value of X, we can use the binomial formula or statistical software. This exercise is very similar to Exercise 13.5 of this Study Guide in which the use of the binomial formula was illustrated. The only difference is that $p = 0.65$ in this exercise and p was 0.3 in Exercise 13.5. The probabilities listed were obtained using the Minitab software.

```
Binomial with n = 5 and p = 0.650000
     x             P(X = x)
   0.00             0.0053
   1.00             0.0488
   2.00             0.1811
   3.00             0.3364
   4.00             0.3124
   5.00             0.1160
```

The probability histogram corresponding to this distribution follows.

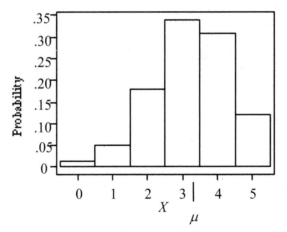

(d) The mean of X is $\mu = np = 5(0.65) = 3.25$ and is indicated on the histogram in part (c). The standard deviation of X is

$$\sigma = \sqrt{np(1-p)} = \sqrt{5(0.65)(0.35)} = 1.067$$

CHAPTER 14

Introduction to Inference

OVERVIEW

Statistical inference provides methods for drawing conclusions about a population from sample data. To make an inference is to make a decision, judgment, conclusion, or estimate about the whole population based on limited information in a sample.

In this chapter, we examine two of the most commonly used methods for making inferences. We focus on making inferences for population means.

(1) Confidence intervals

A **confidence interval** provides an estimate of an unknown parameter of a population or process, along with an indication of how accurate this estimate is and how confident we are that the interval is correct (contains the parameter). Confidence intervals have two parts. One is an interval computed from our data, typically of the form

$$\text{estimate} \pm \text{margin of error}$$

The other part is the **confidence level,** which states the probability that the *method* used to construct the interval will give a correct answer. For example, "95% confidence" means that if you repeatedly collect samples randomly from the same population, using the same methods, and each time constructing a 95% confidence interval based on the sample, then in the long run 95% of these intervals will capture the true value of the parameter you're trying to estimate. Of course, when you apply the method only once, you do not know whether your interval contains the parameter or not. Confidence refers to the long-run proportion of times that that the interval contains the parameter under repeated sampling, not the correctness of any particular interval we compute from one sample.

Suppose we wish to estimate the unknown mean μ of a normal population with known standard deviation σ based on an SRS of size n. A level C confidence interval for μ is

$$\bar{x} \pm z^* \frac{\sigma}{\sqrt{n}}$$

where z^* is such that the probability is C that a standard Normal random variable lies between $-z^*$ and z^* and is obtained from the bottom row in Table C. These z-values are called critical values.

The formula for any specific confidence interval is a recipe that is correct under specific conditions. The most important conditions concern the methods used to produce the data. Many methods (including those discussed here) assume that the data were collected by random sampling. Other conditions, such as the actual distribution of the population, are also important.

(2) Tests of significance

We use confidence intervals when our goal is simply to estimate the value of a parameter. Sometimes, however, the researcher is more interested in testing the plausibility of a claim or statement. For example, a manufacturer might worry if the mean volume of water being dispensed to bottles at a plant exceeds 14.2 ounces. In this case, we look for evidence that there's a problem… namely, that the average volume is more than 14.2 ounces. The emphasis here is not on the question "What are the plausible values of the mean?" but rather on the question "Is the value of interest plausible?" Obviously, these questions are strongly connected and so are the related inferential methods (confidence intervals and tests of significance).

A test of significance is done to assess the evidence against the **null hypothesis** H_0 in favor of an **alternative hypothesis** H_a. Typically, the alternative hypothesis is the effect that the researcher is trying to demonstrate, and the null hypothesis is a statement that the effect is not present. The alternative hypothesis can be either **one-** or **two-sided.**

Tests are usually carried out by first computing a **test statistic.** The test statistic is used to compute a **P-value,** which is the probability of getting a test statistic at least as extreme as the one observed, where the probability is computed when the null hypothesis is true. The P-value provides a measure of how incompatible our data are with the null hypothesis, or how unusual it would be to get data like ours if the null hypothesis were true. Since small P-values indicate data that are unusual or difficult to explain under the null hypothesis, we typically reject the null hypothesis in these cases. In this case, the alternative hypothesis provides a better explanation for our data.

Significance tests of the null hypothesis H_0: $\mu = \mu_0$ with either a one-sided alternative ($H_a : \mu > \mu_0$ or $H_a : \mu < \mu_0$) or two-sided alternative ($H_a : \mu \neq \mu_0$) are based on the test statistic

$$z = \frac{\bar{x} - \mu_0}{\sigma / \sqrt{n}}$$

The use of this test statistic assumes that we have an SRS from a Normal population with known standard deviation σ. When the sample size is large, the assumption of Normality is less critical because the sampling distribution of $\bar{x}$ is approximately Normal. P-values for the test based on z are computed using Table A.

When the P-value is less than a specified value α, we say that the results are **statistically significant at level α,** or we reject the null hypothesis at level α. Tests can alternatively be carried out at a fixed significance level by obtaining the appropriate critical value z^* from the bottom row in Table C.

GUIDED SOLUTIONS

Exercise 14.1

KEY CONCEPTS: Reasoning of statistical estimation

(a) In this problem, we take many samples of size $n = 840$ from a population with standard deviation $\sigma = 60$. In Chapter 11, we studied the sampling distribution of the sample mean, $\bar{x}$. Compute the standard deviation of $\bar{x}$:

Standard deviation of $\bar{x}$ =

(b) Again, we envision repeatedly observing $\bar{x}$, each based on a random sample of 840 men. According to the 68-95-99.7 rule, about 95% of all values of $\bar{x}$ will be within two standard deviations of μ, the unknown mean of $\bar{x}$. That is, 95% of all values of $\bar{x}$ will be within $2 \times$ (standard deviation of $\bar{x}$) of μ. Compute this number:

$2 \times$ (standard deviation of $\bar{x}$) =

(c) Our 95% confidence interval for the population mean score μ based on this one sample is the interval

$\bar{x} - 2 \times$ (standard deviation of $\bar{x}$)

to

$\bar{x} + 2 \times$ (standard deviation of $\bar{x}$)

Exercise 14.5

KEY CONCEPTS: Confidence interval for a population mean; four-step process; stemplot of data to check for departure from Normality; simple assumptions needed for inference

(a) We make a stemplot of these data in order to check for any serious violation of the assumption that the population we're sampling from follows a Normal distribution. The sample stemplot is a quick way to check, and splitting the stems is a way to refine the plot. You can review the details on constructing split stemplots in Chapter 1. Complete the stemplot below:

```
 7 |
 7 |
 8 |
 8 |
 9 |
 9 |
10 |
10 |
11 |
11 |
12 |
12 |
13 |
```

Since this sample was an SRS of 31 girls from a population of all 7th-grade girls in the Midwest school district of discussion, and since the stemplot above doesn't suggest that their distribution is not normal, it is reasonable to say that the simple assumptions required for inference hold here.

(b) The four-step process follows:

State. What is the practical question that requires estimating a parameter?

Plan. Identify the parameter, choose a level of confidence, and select the type of confidence interval that fits your situation.

Solve. Carry out the work in two phases:
      (1) Check the conditions for the interval you plan to use.
      (2) Calculate the confidence interval.
Conclude. Return to the practical question to describe your results in this setting.

To apply the steps to this problem, here are some suggestions:

State. Describe the research question here. Remember that we're trying to learn something about typical IQ test scores among some population.

Plan. We're trying to estimate a population mean, μ. What does μ represent in this problem?

What level of confidence do we want to use to compute our confidence interval?

How will we compute the confidence interval? Write the formula we'll use.

What critical value $z*$ is needed for this confidence interval? You can obtain this from the $z*$ row of Table C.

What is the value of $\bar{x}$?

Solve: Put together the pieces outlined in the *Plan* step above to compute the confidence interval needed. In this problem, $\sigma = 15$ and $n = 31$.

Compute the 99% confidence interval.

Conclude. What does this 99% confidence interval mean? What does it say about average IQ test score?

Exercise 14.6

KEY CONCEPTS: Reasoning behind significance tests

(a) If $\mu = 115$, then scores in the population of older students are Normally distributed, with mean $\mu = 115$ and standard deviation $\sigma = 30$. What is the sampling distribution of $\bar{x}$, the mean of a sample of size $n = 25$? (We studied the sampling distribution of $\bar{x}$ in Chapter 11.) Sketch the density curve of this distribution, making sure to use an appropriate scale.

(b) Mark the two points ($\bar{x} = 118.6$ and $\bar{x} = 125.8$) on your sketch in part (a).

Referring to the sketch, explain in simple language why one result is good evidence that the mean score of all older students is greater than 115 and why the other outcome is not. Think about how far out on the density curve the two points are. You might think about what the 68-95-99.7 rule says about this problem.

Exercise 14.8

KEY CONCEPTS: Stating the null and alternative hypotheses

We're asked to state the null and alternative hypotheses relevant to Exercise 14.6. It is often easiest to state the alternative hypothesis first. The alternative hypothesis is the effect the researcher suspects is true, or is hoping to demonstrate.

What is it about older students that the researcher suspects?

The null hypothesis is the claim our researcher is testing (challenging). What claim is our researcher testing?

Based on these, write the null and alternative hypotheses:

H_0:

H_a:

Exercise 14.15

KEY CONCEPTS: Understanding statistical significance; understanding P-value

In the phrase "significantly more," the word "significant" is a reference to statistical significance. We're saying that the observed difference in infection rates for the two groups (placebo group and vitamin C group) isn't well explained by random chance. That is, if vitamin C really had no impact on incidence of respiratory infection rate, the observed difference in infection rates between the two groups would be very unlikely to occur. Since it did occur, either (1) vitamin C doesn't impact infection rate, and the observed difference was due to random chance, or (2) vitamin C really does reduce infection rates, and that's why we saw such a large difference in the sample groups.

(a) Which of these explanations seems more plausible? Does it seem reasonable to conclude that vitamin C really does reduce respiratory infection rate?

(b) Where does the P-value fit into the discussion above? That is, what does the P-value less than .01 say about the random chance of such a large difference between the group infection rates if vitamin C really has no impact?

Exercise 14.19

KEY CONCEPTS: Significance test for a population mean; two-sided test; four-step process

The four-step process for conducting a test of significance are summarized:

State. What is the practical question that requires a statistical test?

Plan. Identify the parameter, state the null and alternative hypotheses, and choose the type of test that fits your situation.

Solve. Carry out the test in three phases:
   (1) Check the conditions for the test you plan to use.
   (2) Calculate the test statistic.
   (3) Find the P-value.

Conclude. Return to the practical question to describe your results in this setting.

To apply the steps to this problem, here are some suggestions:

State. Describe the research question here.

Plan. The parameter of interest is μ. What does μ represent in this problem?

What does the researcher suspect? What does this say about the alternative hypothesis?
What is the researcher testing (challenging)? What does this say about the null hypothesis?
Write the null and alternative hypotheses:

H_0:

H_a:

Finally, we know that as long as the required "simple conditions" are satisfied, a one-sample z-test is appropriate.

Solve.

(1) Describe what the simple conditions mean in the context of this problem. We'll assume all of the conditions are satisfied, so you don't need to "check" anything. In Chapter 15, we'll examine this issue more carefully.

(2) Calculate the test statistic.

First, compute $\bar{x}$:

Compute the test statistic $z = \dfrac{\bar{x} - \mu_0}{\sigma / \sqrt{n}} =$

(3) Find the *P*-value.
Sketch the Normal density curve that describes the distribution of $\bar{x}$. Mark on it the observed value of z. Shade the area corresponding to the *P*-value. Remember, this is a two-sided test.

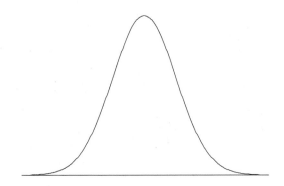

Use Table A to compute the *P*-value:

Conclude. Remember that the smaller your *P*-value is, the *more* evidence you have in favor of the alternative hypothesis. Based on your *P*-value, do we have good evidence that the true conductivity of this liquid differs from 5?

Exercise 14.23

KEY CONCEPTS: Significance from a table; significance level

We observe the sample mean $\bar{x}$ from a random sample of $n = 100$ observations from a population with standard deviation $\sigma = .2887$.

(a) If $\mu = .5$ is true, compute the z statistic corresponding to the test:

$$z = \frac{\bar{x} - \mu_0}{\sigma/\sqrt{n}} =$$

Since the alternative hypothesis is $H_a : \mu \neq 0.5$, this is a two-sided test. To determine whether z is significant at some level α, compare your computed z (ignoring sign) to the value z^* in Table C corresponding to the two-sided P entry for α. Remember that a computed test statistic z is significant at level α if (ignoring sign) it is larger than z^*.

(b) What is the value of z^* corresponding to $\alpha = .05$ for a two-sided test? Ignoring sign, is your z larger or smaller than this z^*? What does this say about the significance of z at level .05?

(c) What is the value of z^* corresponding to $\alpha = .01$ for a two-sided test? Ignoring sign, is your z larger or smaller than this z^*? What does this say about the significance of z at level .01?

(d) Find the two closest values of z^* in Table C between which your computed test statistic z lies:

_____ < z <

What are the corresponding two-sided P-values associated with these values of z^*?

What does this say about the P-value associated with your observed test statistic z?

We'll conclude with a few Exercises that are similar to ones worked earlier. We'll provide slightly less guidance and leave more to you to work out.

Exercise 14.35

KEY CONCEPTS: Confidence interval for a population mean

In this problem, we're asked to give a 90% confidence interval for μ, the average "muscle gap," which describes how much muscle mass men in the population believe they should add in order to be attractive to women.

The researchers took a random sample of $n = 200$ American men and observed a sample average muscle gap of $\bar{x} = 2.35$ kg/m^2. The population they sampled from is Normal with standard deviation $\sigma = 2.5$ kg/m^2.

Write the formula you'll use to compute the confidence interval:

Refer to Table C and obtain the value of z^* corresponding to 90% confidence level:

$z^* =$

Compute the confidence interval:

Provide a brief interpretation of this confidence interval in the context of the problem.

Exercise 14.41

KEY CONCEPTS: Significance test for a population mean

In Exercise 14.35, the parameter μ was described in the context of this problem.

(a) What does the researcher suspect about the value of μ?

Write the null and alternative hypotheses for testing this suspicion:

H_0:

H_a:

(b) In Exercise 14.35, we noted that the researchers took a random sample of $n = 200$ American men and observed a sample average muscle gap of $\bar{x} = 2.35$ kg/m². The population they sampled from is Normal with standard deviation $\sigma = 2.5$ kg/m². Compute the test statistic

(c) Look at the magnitude of your computed test statistic, z. Use the 68/95/99.7 rule to provide some perspective on the meaning of z in this problem.

Exercise 14.42

KEY CONCEPTS: Significance test for a population mean

What is the parameter of interest in this problem, and what does the researcher suspect about it?

(a) Write the null and alternative hypotheses:

H_0:

H_a:

(b) We observed a sample mean of $\bar{x} = 5.29$ based on a sample $n = 148$ male general managers. We assume that the population we sampled from has standard deviation $\sigma = 0.78$. Compute the test statistic.

(c) Compute the *P*-value.

Based on your *P*-value, what would you conclude about the average femininity score for male hotel managers as compared with the average femininity score for men generally (5.19)?

COMPLETE SOLUTIONS

Exercise 14.1

(a) Standard deviation of $\bar{x} = \dfrac{\sigma}{\sqrt{n}} = \dfrac{60}{\sqrt{840}} = 2.07$ points. (Units of measurement are exam points)

(b) 95% of all values of $\bar{x}$ will be within $2 \times$(standard deviation of $\bar{x}$) of μ, or within

$$2 \times (\text{standard deviation of } \bar{x}) = 2(2.07) = 4.14 \text{ points of } \mu$$

(c) Our 95% confidence interval for the population mean score μ based on this one sample is the interval
$\bar{x} - 2 \times$(standard deviation of $\bar{x}$) $= 272 - 4.14 = 267.86$ points
to
$\bar{x} + 2 \times$(standard deviation of $\bar{x}$) $= 272 + 4.14 = 276.14$ points
Hence, we are 95% confident that the mean test score for all men in this population is between 267.86 points and 276.14 points.

Exercise 14.5

(a) A split stemplot of these data:

```
 7 | 2 4
 7 |
 8 |
 8 | 6 9
 9 | 1 3
 9 | 6 8
10 | 0 2 3 3 3 4
10 | 5 7 8
11 | 1 1 2 2 2 4 4 4
11 | 8 9
12 | 0
12 | 8
13 | 0 2
```

This is a sample, and the Normal distribution is only a model. There is no serious departure from Normality indicated here. There's certainly not strong evidence that these data come from a heavily skewed population, so it seems reasonable to conclude that the population of all IQ scores is Normal.

(b) Following the four-step process for confidence intervals,

State. We're interested in estimating the average IQ score for all 7th-grade girls in this Midwest school district.

Plan. In this problem, μ represents the average IQ score for all 7th-grade girls in this Midwest school district. We'll construct a 99% confidence interval for μ.

The formula we'll use is $\bar{x} \pm z^* \dfrac{\sigma}{\sqrt{n}}$. From the z^* row of Table C, for 99% confidence level, $z^* = 2.576$. Finally, $\bar{x} = 105.839$.

Solve. Our 99% confidence interval for μ is:

$$\bar{x} - z^* \frac{\sigma}{\sqrt{n}} = 105.839 - 2.576 \left(\frac{15}{\sqrt{31}} \right) = 105.839 - 6.940 = 98.899$$

to

$$\bar{x} + z^* \frac{\sigma}{\sqrt{n}} = 105.839 + 2.576 \left(\frac{15}{\sqrt{31}} \right) = 105.839 + 6.940 = 112.779$$

Conclude. With 99% confidence, the mean IQ test score for all 7th-grade girls in this Midwest school district is between 98.889 and 112.779.

Exercise 14.6

(a) From Chapter 11, we know that the sampling distribution of $\bar{x}$ is Normal with mean $\mu = 115$ and standard deviation $\sigma = 30/\sqrt{n} = 30/\sqrt{25} = 30/5 = 6$. A sketch of this distribution follows.

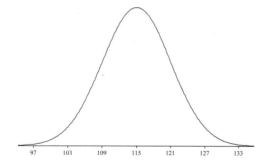

| 97 | 103 | 109 | 115 | 121 | 127 | 133 |

(b) The two points are marked on the following curve.

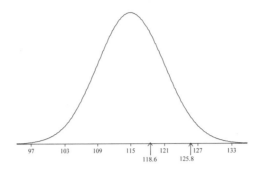

This picture tells us that if for older students $\mu = 115$, a value of $\bar{x} = 125.8$ is much less likely to occur by chance alone than a value of $\bar{x} = 118.6$. In other words, if you actually observed $\bar{x} = 125.8$, this would offer much stronger evidence against the claim that $\mu = 115$ than would an observation of $\bar{x} = 118.6$. That is, $\bar{x} = 125.8$ offers strong evidence that for older students, the mean $\mu > 115$, while $\bar{x} = 118.6$ offers somewhat weaker evidence that $\mu > 115$.

Exercise 14.8

The researcher suspects that for older students, the mean SSHA test score, μ, is more than 115. She's challenging (testing) the claim that the mean test score for older students is 115. Hence,

$$H_0: \mu = 115$$
$$H_a: \mu > 115$$

Exercise 14.15

(a) The researchers have concluded that the observed difference between infection rate in the vitamin C supplemented group and the placebo group is too large to be explained by chance alone. It seems reasonable to conclude that vitamin C is, in fact, reducing infection rate.

(b) The *P*-value measures the probability such a large difference in respiratory infection rates between the two groups would exist if there really is no effect of vitamin C. We see that the *P*-value here is less than .01. In other words, if vitamin C really has no impact, the probability the placebo group would have so many more infections than the vitamin C group by random chance is less than .01.

Exercise 14.19

State. The researcher wonders if the true conductivity of the liquid differs from 5.
Plan. In this problem, μ represents the true conductivity of the liquid - the average value of conductivity as measured by an accurate measuring device. The researcher wonders if μ differs from 5. It could be larger or smaller, so this is a two-sided test. The hypotheses are
$$H_0: \mu = 5$$
$$H_a: \mu \neq 5$$

Solve. In this chapter, we assume the simple conditions are true. For illustration, we outline them here: First, we assume that the sample of six measurements of conductivity of this liquid represents a simple random sample from a population of all such measurements. We assume this population distribution of all measurements is Normal. We're told that $\sigma = 0.2$.

Now, $\bar{x} = (5.32 + 4.88 + 5.10 + 4.73 + 5.15 + 4.75)/6 = 4.9883$

Hence, the test statistic $z = \dfrac{\bar{x} - \mu_0}{\sigma/\sqrt{n}} = \dfrac{4.9883 - 5}{0.2/\sqrt{6}} = -0.14$.

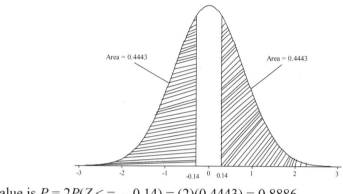

Our *P*-value is $P = 2P(Z <= -0.14) = (2)(0.4443) = 0.8886$.

Conclude. If we were to repeatedly randomly select six measurements of conductivity from a liquid with true conductivity 5, then more than 88% of the resulting sample means would be at least as far from 5 as our observed sample mean ($\bar{x} = 4.9883$). In other words, our observed $\bar{x}$ is consistent with the assumption that $\mu = 5$, and offers little evidence that the true conductivity of this liquid differs from 5.

Exercise 14.23

(a) $z = \dfrac{\bar{x} - \mu_0}{\sigma/\sqrt{n}} = \dfrac{0.4365 - 0.5}{.2887/\sqrt{100}} = -2.20$

Since this is a two-sided test, refer to the Table C row labeled "Two-sided *P*". Ignoring the sign of *z*,

(b) $z = 2.20$ is larger than $z^* = 1.96$. Hence, it is significant at the 5% level ($\alpha = .05$).

(c) $z = 2.20$ is smaller than $z^* = 2.576$. Hence, it is *not* significant at the 1% level ($\alpha = .01$).

(d) Note that $z = 2.20$ is between the two consecutive Table C entries of $z^* = 2.054$ (which corresponds to a test at the 4% level) and $z^* = 2.326$ (which corresponds to a test at the 2% level). We can say that the test is significant at the 4% level ($\alpha = .04$), but not at the 2% level ($\alpha = .02$). Hence, the *P*-value lies between .02 and .04. This provides strong evidence against the assumption that $\mu = .5$ and in support of the conclusion that $\mu \neq .5$.

Exercise 14.35

According to Table C, $z^* = 1.645$. Our 90% confidence interval for μ is

$$\bar{x} - z^* \frac{\sigma}{\sqrt{n}} = 2.35 - 1.645\left(\frac{2.5}{\sqrt{200}}\right) = 2.35 - 0.29 = 2.06 \text{ kg/m}^2$$

to

$$\bar{x} + z^* \frac{\sigma}{\sqrt{n}} = 2.35 + 1.645\left(\frac{2.5}{\sqrt{200}}\right) = 2.35 + 0.29 = 2.64 \text{ kg/m}^2$$

With 90% confidence, the mean "muscle gap" for young men is between 2.06 kg/m² and 2.64 kg/m².

Exercise 14.41

(a) If μ denotes the average "muscle gap" for young American men, the researcher suspects that $\mu > 0$.

$$H_0: \mu = 0$$
$$H_a: \mu > 0$$

(b) $z = \dfrac{\bar{x} - \mu_0}{\sigma/\sqrt{n}} = \dfrac{2.35 - 0}{2.5/\sqrt{200}} = 13.29$

(c) Under the null hypothesis, a sample mean like ours (2.35) would be more than 13 standard deviations away from where we expected it to be (0). According to the 68/95/99.7 rule, values of z more than 3 are very rare. It seems that under the null hypothesis, our observed sample mean should not have occurred, and the corresponding P-value is very small (effectively zero!). This constitutes overwhelming evidence in support of the researcher's suspicion that young Americans do have a (positive) "muscle gap" ... they feel they should become more muscular in order to be attractive to women.

Exercise 14.42

(a) The parameter m denotes the average femininity score for male hotel managers. The researcher suspects that m is different from 5.19, which is the average femininity score for the general population of all men. Hence, the hypotheses are $H_o: \mu = 5.19$ versus $H_a: \mu \neq 5.19$.

(b) We have a sample of $n = 148$ male hotel managers. They tested an average femininity score of 5.29. We're assuming that the standard deviation for male hotel managers is the same as for the general population of all men, $\sigma = 0.78$. The test statistic is

$$z = \frac{\bar{x} - \mu_0}{\sigma/\sqrt{n}} = \frac{5.29 - 5.19}{0.78/\sqrt{148}} = 1.56$$

{This means that the observed sample mean (5.29) is 1.56 standard deviations above the value we would have anticipated (5.19) if the null hypothesis is correct.}

(c) P-value $= 2P(Z > 1.56) = 2(1 - 0.9406) = 2(0.0594) = 0.1188$. This provides only fairly weak evidence in support of the researcher's suspicion. That is, we don't have much evidence to support a claim that male hotel managers differ in their mean femininity score from that of males generally.

CHAPTER 15

THINKING ABOUT INFERENCE

OVERVIEW

In Chapter 14, two methods for making inferences about a population mean were described. Inference based on a confidence interval or significance test can be trusted only under specific conditions: The data are a simple random sample; the population being sampled from has a Normal distribution; and the population standard deviation σ is known.

In practice, effective use of these procedures requires judgment on the part of the practitioner. In fact, the population standard deviation σ is rarely known. Later we'll study how to deal with the situation in which σ is not known using methods based on these. The other two conditions are more subjectively assessed. What's most important for any statistical procedure is that the data come from random sampling of some sort, such as a simple random sample or a randomized comparative experiment. In practice, it's sometimes the case that the data were not really selected randomly. Even so, one can sometimes act as if the sample is a simple random sample. This is true if there are no problems such as nonresponse in samples. Finally, although we assume that the population we're sampling from is Normal in distribution, this assumption is less important. This is because the z procedures are based on the Normality of the sample mean, $\overline{x}$, rather than the Normality of the population. The central limit theorem says that the distribution of $\overline{x}$ is more Normal in distribution than the individual members of the population, and it becomes more Normal as the sample size increases.

The z procedures described in Chapter 14 are based on $\overline{x}$, so they are sensitive to outliers. The presence of outliers may suggest that the population we're sampling from is far from Normal. In this case, we recommend other procedures for inference about the population mean (see Chapter 25 on your text CD).

Suppose we wish to estimate the unknown mean μ of a normal population with known standard deviation σ based on an SRS of size n. The level C confidence interval for μ is $\overline{x} \pm z^* \sigma / \sqrt{n}$. The margin of error, $z^* \sigma / \sqrt{n}$ decreases when

- The confidence level C decreases
- The sample size n increases
- The population standard deviation σ decreases.

If we our confidence interval to have margin of error no greater than *m*, the sample size needed is $n = \left(\dfrac{z^* \sigma}{m} \right)^2$. Of course, the margin of error only accounts for chance variation due to sampling. Other problems such as nonresponse or undercoverage are often more problematic.

When describing the outcome of a hypothesis test, it is more informative to give the *P*-value than to just reject or not reject a decision at a particular significance level α. The traditional levels of 0.01, 0.05, and 0.10 are arbitrary and serve as rough guidelines. Researchers often use different levels of significance depending on the plausibility of the null hypothesis and the consequences of rejecting the null hypothesis. There is no sharp boundary between significant and insignificant, only increasingly strong evidence as the *P*-value decreases.

When testing hypotheses with a very large sample, the *P*-value can be very small for effects that may not be of interest. Don't confuse small *P*-values with large or important effects. Statistical significance is not the same as practical significance. Plot the data to display the effect you are trying to show and also give a confidence interval that says something about the size of the effect.

Just because a test is not statistically significant doesn't imply that the null hypothesis is true. Statistical significance may occur when the test is based on a small sample size. Finally, if you run enough tests, you will invariably find statistical significance for one of them. Be careful in interpreting the results when testing many hypotheses on the same data.

From the point of view of making decisions, H_0 and H_a are just two statements of equal status that we must decide between. One chooses a rule for deciding between H_0 and H_a on the basis of the probabilities of the two types of errors we can make. A **Type I error** occurs if H_0 is rejected when it is in fact true. A **Type II error** occurs if H_0 is accepted when in fact H_a is true. There is a relation between α-level significance tests and testing from the decision-making point of view. The probability of Type I error is α.

To compute the Type II error probability of a significance test about a mean of a normal population:

- Write the rule for accepting the null hypothesis in terms of $\bar{x}$.
- Calculate the probability of accepting the null hypothesis when the alternative is true.

The **power** of a significance test is always calculated at a specific alternative hypothesis and is the probability that the test will reject H_0 when that alternative is true. The power of a test against any particular alternative is 1 minus the probability of a Type II error. Power is usually interpreted as the ability of a test to detect an alternative hypothesis or as the sensitivity of a test to an alternative hypothesis. The power of a test can be increased by increasing the sample size as the significance level remains fixed.

GUIDED SOLUTIONS

Exercise 15.1

KEY CONCEPTS: Conditions for inference in practice

In this problem, we are asked to compare and comment on three reasons for which a confidence interval might not be useful. There are a few ways for a confidence interval to be of little use. Most importantly, the interval is of little use if it can't be trusted (it isn't valid). Remember that for the z procedures are trustworthy only if the following conditions are valid: (1) Data are a simple random sample; (2) the population being sampled from has a Normal distribution; and (3) the population standard deviation σ is known. Of course, these conditions rarely hold precisely, but sometimes we're able to act as though they hold approximately. It's also true that the interval will be of little use if the margin of error is very large relative to what we're estimating.

(a) Is it a problem that the course is small, so that the margin of error will be large?

(b) If many in the class do refuse to respond, is the confidence interval untrustworthy?

(c) If the students in the course can't be considered a random sample from the population of all college-age adults, how does this impact the trustworthiness of the confidence interval?

Which of these three reasons is most problematic?

Exercise 15.5

KEY CONCEPTS: How confidence intervals behave

In Example 14.1 of Chapter 14, a 95% confidence interval for the average BMI of women aged 20 to 29 is given by 26.2 to 27.4. This was based on a random sample of 654 women, sampled from a population with standard deviation $\sigma = 7.5$. The margin of error associated with this confidence interval is ± 0.6. The observed sample mean is $\bar{x} = 26.8$.

The margin of error associated with a confidence interval is $\pm z^* \dfrac{\sigma}{\sqrt{n}}$.

(a) Compute the margin of error as described in Example 14.1, but assume that the sample size was 100.

(b) Now, compute the margin of error again, but assume that the sample size was 400.

Finally, compute the margin of error again, but assume the sample size was 1600.

(c) When we increase the sample size, what happens to the size of the margin of error (keeping confidence level and population standard deviation the same)?

Exercise 15.6

KEY CONCEPTS: Sources of error included in margin of error

Here, Gallup is estimating the percentage of Americans that are confident that the food available at most grocery stores is safe to eat. Based on a presumably random sample, Gallup's margin of error associated with a 95% confidence interval for this unknown percentage is $\pm 3\%$. In the context of this problem, this means that if we repeated the survey many times, in the long run our sample percentage would be within about 3% of the true percentage 95% of the time.

(a) Does the margin of error account for systematically missing a group of people (the ones with no land line)?

(b) Does the margin of error account for people that don't participate or cannot be reached? What if people that can't be reached feel differently about food safety than people that can be reached?

(c) Does the margin of error account for the fact that the sample percentage (the percentage of the sample confident in food safety) will randomly vary under repeated sampling?

Exercise 15.7

KEY CONCEPTS: Statistical significance and practical significance

We need to carry out a test of H_0: $\mu = 518$, H_a: $\mu > 518$ based on a random sample of $n = 50$ students and observe a sample mean $\bar{x}$. The population standard deviation is $\sigma = 114$.

(a) In this case, $\bar{x} = 544$.

Compute the test statistic: $\qquad Z = \dfrac{\bar{x} - \mu_0}{\sigma / \sqrt{n}} =$

Compute the *P*-value:

Is this result significant at the 5% level?

(b) In this case, $\bar{x} = 545$.

Compute the test statistic: $Z = \dfrac{\bar{x} - \mu_0}{\sigma/\sqrt{n}} =$

Compute the *P*-value:

Is this result significant at the 5% level?

Exercise 15.10

KEY CONCEPTS: Multiple analyses

(a) Suppose you test a single subject for ESP using the 1% level. This means that, even if the subject does not have ESP ability, our test would falsely lead us to conclude that he/she does with probability .01.

Now, if all 500 subjects are simply guessing randomly, how many would you expect to achieve a score that has such a *P*-value (< 0.01)?

(b) If 500 people are tested for ESP, we expect some of them to pass the test even if they're guessing. So if a handful of them pass, we don't know whether they actually have ESP ability or if they passed by guessing. What would you suggest the researcher do now to make this determination?

Exercise 15.11

KEY CONCEPTS: Sample size for confidence interval

Given the Normal population standard deviation σ, the sample size required to construct a confidence interval with specified margin of error m is

$$n = \left(\frac{z^* \sigma}{m} \right)^2$$

In this problem, we're to construct a 95% confidence interval with margin of error ± 1. We know $\sigma = 7.5$.

Compute the sample size required:

Exercise 15.13

KEY CONCEPTS: Power of a statistical test

We're testing the hypotheses H_0: $\mu = 5$, H_a: $\mu \neq 5$. A significance test's power is its ability to recognize an effect that is present. That is, if the conductivity of a liquid is 5.1, the correct decision would be to reject H_0. The power of a significance test is the chance a sample will lead us to reject H_0.

Suppose we repeatedly measure this liquid six times, each time using the sample to test these hypotheses.

a) What does "power = 0.23" mean?

b) If the test correctly concludes that conductivity differs from 5 only 23% of the time, how often does it fail to recognize this? What does this mean about the test's ability to "protect" you against a liquid with conductivity 5.1?

Exercise 15.14

KEY CONCEPTS: Power of a statistical test; factors that influence the power of a statistical test

In the setting of Exercise 15.13, power represents the probability that a sample will lead us to correctly conclude that a liquid with real conductivity 5.1 does not have conductivity 5.

(a) Will this probability increase if we collect more measurements or fewer measurements?

(b) The level of significance of a test is a (subjectively selected) threshold for determining that a sample is significant. By using a higher level of significance, we reject H_0 with greater ease because more random samples will yield a P-value below the higher level of significance.

Will the probability of correctly rejecting H_0 increase if we increase the level of significance from $\alpha = .05$ to $\alpha = .10$?

(c) The farther μ is from the value specified by the null hypothesis, the more likely it is to obtain a sample that is significant.

Will the probability of correctly rejecting H_0 increase if we shift interest to the alternative $\mu = 5.2$?

Exercise 15.17

KEY CONCEPTS: Type I and Type II error probabilities

(a) Write the two hypotheses. Remember, we usually take the null hypothesis to be the statement of "no effect."

H_0:

H_a:

Describe the two types of errors as "false positive" and "false negative" test results.

(b) Which error probability would you choose to make smaller? Why?

Exercise 15.52

KEY CONCEPTS: Computing Type I and Type II error probabilities

A random sample of $n = 9$ is selected from a Normal population with mean μ and standard deviation $\sigma = 1$. Recall that the sampling distribution of $\bar{x}$ is Normal with mean m and standard deviation $\sigma/\sqrt{n}$.

The hypotheses being tested are
$$H_0 : \mu = 0$$
$$H_a : \mu > 0$$
The researcher will reject H_0 if $\bar{x} > 0$.

First, if H_0 is true, what is the sampling distribution of $\bar{x}$?

If $\mu = 0.1$, what is the sampling distribution of $\bar{x}$?

If $\mu = 0.1$, what is the sampling distribution of $\bar{x}$?

(a) A Type I error occurs if we reject H_0 when H_0 is true. The researcher will reject H_0 if $\bar{x} > 0$. What is the probability that we reject H_0 when H_0 is true?

A Type II error occurs if we fail to reject H_0 when H_0 is false. The researcher fails to reject H_0 if $\bar{x} \leq 0$.

(b) What is the probability of that we reject H_0 when H_0 is false and $\mu = 0.3$?

(c) What is the probability of that we reject H_0 when H_0 is false and $\mu = 1$?

COMPLETE SOLUTIONS

Exercise 15.1

(a) It's true that margin of error decreases as the sample size increases. It's difficult to say how large the margin of error would be in this case, but we might speculate that (1) student's ratings are likely to cluster about a common high value because the movie is a hit. Hence, the standard deviation of their ratings (not provided with this problem) is likely to be small, and/or (2) the sample of 25 students is not terribly small. Both of these would act to hold the margin of error down. At any rate, a large margin of error may diminish the usefulness of a confidence interval, but it has no bearing on the interval's trustworthiness.

(b) It's a problem if students that refuse to respond tend to feel differently about the movie than those that respond. This problem, called nonresponse, is especially problematic for survey questions involving controversial topics. It's not obvious here that nonresponse would be a serious problem, as students taking a class on filmmaking are likely to be willing to rate the movie being reviewed.

(c) There's never any way to overcome the problem of a nonrepresentative sample. In this case, the instructor is using students in a class on filmmaking as a sample of all college-age adults. Clearly this sample can hardly represent this population.

While the problems in (a.) and (b) are not necessarily issues we can ignore, the problem mentioned in (c) is easily the biggest reason the professor's confidence interval is of little use.

Note: The comments in (a) and (b) above involve some speculation, but the main point of this problem is that problems that threaten the trustworthiness of a procedure usually trump problems that induce random error as in (a) or bias as in (b) in an estimate.

Exercise 15.5

(a) If 100 women were sampled, the margin of error would be $1.96\frac{7.5}{\sqrt{100}} = 1.47$.

(b) If 400 women were sampled, the margin of error would be $1.96\frac{7.5}{\sqrt{400}} = 0.735$.

If 1600 women were sampled, the margin of error would be $1.96\frac{7.5}{\sqrt{1600}} = 0.3675$.

(c) As the sample size increases, the margin of error decreases. In fact, if you look closely, notice that every time we quadruple the sample size, the margin of error is halved.

Exercise 15.6

(a) Suppose people that don't have a land line feel very differently about the safety of food at grocery stores than people that do have a land line. Then, Gallup's estimate of the percentage of all people that feel this food is safe will be biased. The margin of error measures random error, not systematic error such as bias. Hence, this sort of error is not included (accounted for) in the margin of error.

(b) If many people can't be reached or refuse to participate, this may or may not cause the sample to be nonrepresentative of all people. It will be nonrepresentative of all people if the people that respond feel differently than people that don't respond. Otherwise, nonresponse won't be a problem. Either way, as in (a), this sort of problem isn't accounted for in the margin of error.

(c) The margin of error measures how much the estimate will vary in repeated random samples. The very reason estimates vary is chance variation in the random selection of individuals for the survey. Hence, this source of error is precisely the sort accounted for by the margin of error.

Exercise 15.7

(a) $Z = \dfrac{544 - 518}{114/\sqrt{50}} = 1.61$. Hence, the P-value is $P = P(Z \geq 1.61) = 1 - 0.9463 = 0.0537$.

This result is not significant at the 5% level. Technically, at the 5% level, we don't quite have enough evidence to conclude that students that undergo this rigorous training improve their SAT score, on average.

(b) $Z = \dfrac{545 - 518}{114/\sqrt{50}} = 1.67$. Hence, the P-value is $P = P(Z \geq 1.67) = 1 - 0.9525 = 0.0475$.

This result is (barely) significant at the 5% level. Technically, at the 5% level, we have enough evidence to conclude that students that undergo this rigorous training improve their SAT score, on average.

Of course, the two observed sample means (544 and 545) are so close that nobody would view their difference as practically important. In both cases, there is some evidence of the program's success at raising average SAT score. It just so happens that in case (b), there's barely enough evidence to meet the subjectively selected 5% level threshold; while in case (a), the evidence falls just short of reaching that threshold. But the threshold itself is totally subjective.

Exercise 15.10

(a) A P-value of 0.01 means that the probability a subject would do so well when merely guessing is only 0.01. Among 500 subjects, all of whom are merely guessing, we would therefore expect 1%, or 5, of them to do significantly better than random guessing ($P < 0.01$). Thus, in 500 tests, it is not unusual to see four results with P-values on the order of 0.01, even if all are guessing and none have ESP.

(b) These four subjects only should be retested with a new, well-designed test. If all four again have low P-values (say, below 0.01 or 0.05), we have real evidence that they are not merely guessing. In fact, if any one of the subjects has a very low P-value (say, below 0.01), it would also be reasonably compelling evidence that the individual is not merely guessing. A single P-value on the order of 0.10, however, would not be particularly convincing.

Exercise 15.11

The required sample size is $n = \left(\dfrac{1.96 \times 7.5}{1}\right)^2 = 216.09$. Round this up to 217 individuals.

Exercise 15.13

(a) "power = 0.23" means that if the true conductivity of the liquid is really 5.1, and we repeatedly test these hypotheses at the 5% level (each time based on a new random sample of 6 measurements), we'll correctly conclude that the conductivity is different from 5 about 23% of the time.

(b) Our testing procedure (taking six measurements) will lead us to conclude that the true conductivity of the liquid is different from 5 only 23% of the time, when the true conductivity is really 5.1. That is, when the conductivity is really 5.1, we'll not recognize that it's different from 5 77% of the time.

Exercise 15.14

(a) If we make a larger number of measurements, power will increase.

(b) If we use $\alpha = .10$ instead of $\alpha = .05$, we conclude that the sample is statistically significant more easily (often). For example, what if the P-value is $P = .07$? We would conclude significance at the level $\alpha = .10$, but not at the level $\alpha = .05$. By making it easier to conclude significance, we make it easier to conclude that the conductivity is different from 5. Hence, power will increase.

(c) If we shift our interest to the alternative $\mu = 5.2$, power will increase. This is because if $\mu = 5.2$, it will become more likely to obtain a sample mean that is deemed significantly different from 5.

Exercise 15.17

(a) The two hypotheses are

H_0: the patient has no medical problem
H_a: the patient has a medical problem

One possible error is to decide the patient has a medical problem (and send them to the doctor) when, in fact, the patient does not really have a medical problem. This is a Type I error and in this setting could be called a false positive.

The other type of error is to decide the patient has no medical problem when, in fact, the patient does have a problem. This is a Type II error and in this setting could be called a false negative.

(b) Most would say that a Type II error is most problematic in this setting because failing to recognize a real medical problem seems worse than falsely diagnosing one. To avoid this, we'll choose to decrease the probability for a Type II error.

Exercise 15.52

(a) The probability of a Type 1 error is the probability that we reject H_0 when it is true. We reject H_0 when $\bar{x} > 0$. Since the sampling distribution is symmetric about $\mu = 0$ under H_0, $P(\text{Type 1 error}) = P(\bar{x} > 0$ when $\mu = 0) = 0.50$.

(b) If $\mu = 0.3$, the sampling distribution of $\bar{x}$ is Normal with mean $\mu = 0.3$ and standard deviation $\sigma/\sqrt{n} = 1/\sqrt{9} = 1/3$. A Type II error occurs if we fail to reject H_0 when we should reject H_0. In this case, $P(\text{Type II error}) = P(\bar{x} \leq 0$ when $\mu = 0.3) = P(z \leq \frac{0 - 0.3}{1/3}) = P(Z \leq -0.90) = 0.1841$.

(c) If $\mu = 1$, the sampling distribution of $\bar{x}$ is Normal with mean $\mu = 1$ and standard deviation $\sigma/\sqrt{n} = 1/\sqrt{9} = 1/3$. A Type II error occurs if we fail to reject H_0 when we should reject H_0. In this case, $P(\text{Type II error}) = P(\bar{x} \leq 0 \text{ when } \mu = 1) = P(Z \leq \frac{0-1}{1/3}) = P(Z \leq -3) = 0.0013$.

CHAPTER 16

FROM EXPLORATION TO INFERENCE: PART II REVIEW

To assist you in reviewing the material in Chapters 8–16, we provide the text chapter and related problems in this Study Guide for each of the odd-numbered review exercises. Other than pointing you in the right direction, we provide no additional hints or solutions. At this point, you should be able to work these problems on your own with minimal assistance. As a final challenge, we encourage you to work some of the Supplementary Exercises, which integrate more fully the material in these chapters.

Exercise 16.1
Text Location – Chapter 9 for observational studies and experiments
Related Study Guide exercises – Exercises 9.1, 9.5, 9.12

Exercise 16.3
Text Location – Chapter 8 for selecting an SRS
Related Study Guide exercises – Exercise 8.7

Exercise 16.5
Text Location – Chapter 9 for designing an experiment, randomization, and response variable
Related Study Guide exercises – Exercises 9.1, 9.9, 9.37

Exercise 16.7
Text Location – Chapter 9 for designing an experiment, randomization, and response variable
Related Study Guide exercise – Exercise Exercises 9.1, 9.9, 9.37

Exercise 16.9
Text Location – Chapter 8 for identifying bias in samples
Related Study Guide exercises – Exercises 8.13, 8.37

Exercise 16.11
Text Location – Chapter 14 for stating hypotheses in a test of hypotheses
Related Study Guide exercises – Exercises 14.8, 14.19, 14.41, 14.42

Exercise 16.13
Text Location – Chapter 14 for confidence interval for a population mean
Related Study Guide exercises – Exercises 14.1, 14.5, 14.35

Exercise 16.15
Text Location – Chapter 15 for sample size requirement
Related Study Guide exercises – Exercise 15.11

Exercise 16.17
Text Location – Chapter 14 for confidence interval for a population mean
Related Study Guide exercises – Exercises 14.1, 14.5, 14.35

Exercise 16.19
Text Location – Chapter 14 for confidence interval for a population mean; Chapter 15 for how
            confidence intervals behave
Related Study Guide exercises –Exercises 14.1, 14.5, 14.35, 15.5

Exercise 16.21
Text Location – Chapter 14 for significance test for a population mean
Related Study Guide exercises – Exercise 14.19, 14.41, 14.42

Exercise 16.23
Text Location – Chapter 10 for identifying a sample space
Related Study Guide exercises – Exercise 10.5

Exercise 16.25
Text Location – Chapter 10 for discrete probability model
Related Study Guide exercises – Exercise 10.36

Exercise 16.27
Text Location – Chapter 11 for sampling distribution of the sample mean
Related Study Guide exercises – Exercises 11.13, 11.27, 11.38, 11.40

Exercise 16.29
Text Location – Chapter 14 for interpreting a P-value as a measure of significance
Related Study Guide exercises – Exercises 14.23, 14.41, 14.42

Exercise 16.31
Text Location – Chapter 5 for interpreting r^2; Chapter 14 for measuring significance with a P-value
Related Study Guide exercises – Exercises 5.32, 14.19, 14.23

CHAPTER 17

INFERENCE ABOUT A POPULATION MEAN

OVERVIEW

Confidence intervals and significance tests for the mean μ of a normal population are based on the sample mean $\bar{x}$ of an SRS. When the sample size n is large, the central limit theorem suggests that these procedures are approximately correct for other population distributions. In Chapter 14 of your text, the (unrealistic) situation is considered in which we know the population standard deviation, σ. In this chapter, we consider the more realistic case where σ is not known and we must estimate σ from our SRS by the sample standard deviation s. In Chapter 14 we used the **one-sample z statistic**

$$z = \frac{\bar{x} - \mu}{\sigma/\sqrt{n}}$$

which has the $N(0,1)$ distribution. Replacing σ with s, we now use the **one-sample t statistic**

$$t = \frac{\bar{x} - \mu}{s/\sqrt{n}}$$

which has the **t distribution** with $n - 1$ **degrees of freedom.** For every positive value of k there is a t distribution with k degrees of freedom, denoted $t(k)$. All are symmetric, bell-shaped distributions, similar in shape to normal distributions but with greater spread. As k increases, $t(k)$ approaches the $N(0,1)$ distribution.

A level C **confidence interval for the mean** μ of a normal population when σ is unknown is

$$\bar{x} \pm t^* \frac{s}{\sqrt{n}}$$

where t^* is the upper $(1 - C)/2$ critical value of the $t(n - 1)$ distribution, whose value can be found in Table C in your text or from statistical software. The one-sample t confidence interval has the form estimate $\pm\, t^*\text{SE}_{\text{estimate}}$, where "SE" stands for **standard error.**

Significance tests of H_0: $\mu = \mu_0$ are based on the one-sample t statistic. P-values or fixed significance levels are computed from the $t(n - 1)$ distribution using Table C or, more commonly in practice, using statistical software.

One application of these one-sample *t* procedures is to the analysis of data from **matched pairs** studies. We compute the differences between the two values of a matched pair (often before and after measurements on the same unit) to produce a single sample value. The sample mean and standard deviation of these differences are computed. Depending on whether we are interested in a confidence interval or a test of significance concerning the difference in the population means of matched pairs, we use either the one-sample confidence interval or the one-sample significance test based on the *t* statistic.

For larger sample sizes, the *t* procedures are fairly **robust** against nonnormal populations. As a rule of thumb, *t* procedures are useful for nonnormal data when $n \geq 15$, unless the data show outliers or strong skewness. For samples of size $n \geq 40$, *t* procedures can be used for even clearly skewed distributions. For smaller samples, it is a good idea to examine stemplots or histograms before you use the *t* procedures to check for outliers or skewness.

GUIDED SOLUTIONS

Exercise 17.7

KEY CONCEPTS: One-sample *t* confidence intervals, checking assumptions

The four-step process follows.

State. What is the practical question that requires estimating a parameter?

Plan. Identify the parameter, choose a confidence level, and select the appropriate interval.

Solve. Check the conditions and calculate the confidence interval.

Conclude. Return to the practical question to describe your results in this setting.

To apply the steps to this problem, here are some suggestions.

State. What characteristic of ancient air is of interest here? What question about this characteristic do we wish to answer?

Plan. What confidence interval method will we use? What is the level of confidence?

Solve. Are the conditions for inference satisfied? (Do we have an SRS? Is the population approximately normal?) With a sample size of only $n = 9$, the most sensible graph for determining whether the population is approximately normal is probably a stemplot. Complete the stemplot that follows. Use split stems and use just the numbers to the left of the decimal place.

```
4 |
5 |
5 |
6 |
6 |
```

What do you conclude?

To compute a level C confidence interval, we use the formula $\bar{x} \pm t^* \dfrac{s}{\sqrt{n}}$, where t^* is the upper $(1 - C)/2$ critical value of the $t(n - 1)$ distribution, which can be found in Table C. Fill in the missing values. Don't forget to subtract 1 from the sample size when finding the appropriate degrees of freedom for the t confidence interval.

$C =$
$n =$
$t^* =$

Now compute the values of $\bar{x}$ and s from the data given. Use statistical software or a calculator.

$\bar{x} =$ $s =$

Substitute all these values into the formula to complete the computation of the 95% confidence interval.

$$\bar{x} \pm t^* \dfrac{s}{\sqrt{n}} =$$

Conclude. State clearly what you have found in terms of the mean percent of nitrogen in ancient air.

Exercise 17.28

KEY CONCEPTS: Confidence intervals based on the one-sample t statistic, assumptions underlying t procedures

(a) To compute a level C confidence interval, we use the formula $\bar{x} \pm t^* \dfrac{s}{\sqrt{n}}$, where t^* is the upper $(1 - C)/2$ critical value of the $t(n - 1)$ distribution, which can be found in Table C. Fill in the missing values. Don't forget to subtract 1 from the sample size when finding the appropriate degrees of freedom for the t confidence interval.

$C =$
$n =$
$t^* =$

The values of $\bar{x}$ and s are given in the problem.

$\bar{x} =$ $s =$

Substitute all these values into the formula to complete the computation of the 95% confidence interval.

$$\bar{x} \pm t^* \frac{s}{\sqrt{n}} =$$

(b) What are the assumptions required for the t confidence interval? Which assumptions are satisfied and which may not be? How were the subjects in the study obtained? How were the subjects in the placebo group obtained?

Exercise 17.45

KEY CONCEPTS: Matched pairs experiments, one-sample t tests

(a) This is a matched pairs experiment. The matched pair of observations are the right-hand and left-hand times on each subject. To avoid confounding with time of day, we would probably want subjects to use both knobs in the same session. We would also want to randomize which knob the subject uses first. How might you do this randomization? What about the order in which the subjects are tested?

(b) The four-step process follows.

   State. What is the practical question that requires a statistical test?

   Plan. Identify the parameter, state null and alternative hypotheses, and choose the appropriate test.

   Solve. (1) Check the conditions, (2) calculate the test statistic, and (3) find the P-value.

   Conclude. Return to the practical question to describe your results in this setting.

To apply the steps to this problem, here are some suggestions.

State. What characteristic of the experiment is of interest here? What question about this characteristic do we wish to answer?

Plan. The project hopes to show that right-handed people find right-hand threads easier to use than left-hand threads. In terms of the mean μ for the population of differences

$$\text{(left thread time)} - \text{(right thread time)}$$

what do we wish to show? This hypothesis would be the alternative. What are H_0 and H_a (in terms of μ)?

H_0:

H_a:

What statistical test will you use for to test these hypotheses?

Solve. Are the conditions for inference satisfied? (Was the experiment properly randomized? Is the condition of normality satisfied?) For data from a matched pairs study, we compute the differences between the two values of a matched pair to produce a single sample value. These differences are as follows.

Right thread	Left thread	Difference = Left – Right
113	137	24
105	105	0
130	133	3
101	108	7
138	115	−23
118	170	52
87	103	16
116	145	29
75	78	3
96	107	11
122	84	−38
103	148	45
116	147	31
107	87	−20
118	166	48
103	146	43
111	123	12
104	135	31
111	112	1
89	93	4
78	76	−2
100	116	16
89	78	−11
85	101	16
88	123	35

Use the axes below to make a histogram of the differences. Use as class intervals –40 through –20, –20 through 0, and so on.

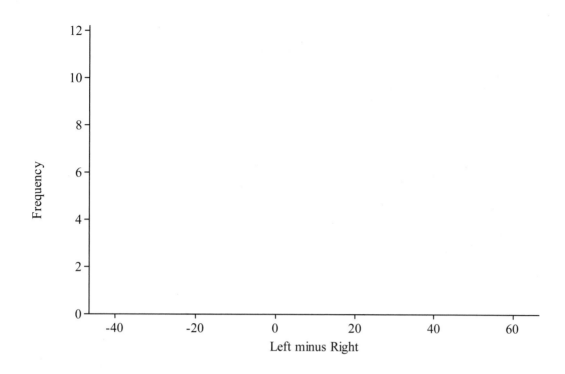

Does the normality condition appear to be satisfied?

The sample mean and the standard deviation of these differences need to be computed. Fill in their values. Use statistical software or a calculator.

$\bar{x} =$ $s =$

Now use the one-sample significance test based on the t statistic. What value of μ_0 should be used?

$$t = \frac{\bar{x} - \mu_0}{s/\sqrt{n}} =$$

From the value of the t statistic and Table C (or using statistical software), the P-value can be computed. Using Table C, between what two values does the P-value lie?

$\leq P\text{-value} \leq$

Exact P-value from software $=$

Note: This problem is most easily done directly using statistical software. The software will compute the differences, the t statistic, and the P-value. Consult your user manual to see how to do one-sample t tests.

Conclude. State clearly what you have found in terms of mean time to move the indicator a fixed distance. Relate this to the original goal of the project, namely to show that right-handed people find right-hand threads easier to use than left-hand threads.

Exercise 17.47

KEY CONCEPTS: Matched pairs experiments, confidence intervals

Taking the 25 differences (left – right), we get the mean and standard deviation of the differences as $\bar{x} = 13.32$, $s = 22.94$ (see Exercise 17.45 in this Study Guide). To compute a level C confidence interval, use the formula

$$\bar{x} \pm t^* \frac{s}{\sqrt{n}} =$$

where t^* is the upper $(1 - C)/2$ critical value of the $t(n - 1)$ distribution, which can be found in Table C. Substitute all these values into the formula to complete the computation of the 95% confidence interval. Don't forget to subtract one from the sample size when finding the appropriate degrees of freedom for the t confidence interval.

As an alternative to computing the mean of the differences, you could evaluate the ratio of the mean time for right-hand threads as a percent of left-hand threads to help determine whether the time saved is of practical importance.

$$\bar{x}_R / \bar{x}_L =$$

COMPLETE SOLUTIONS

Exercise 17.7

State. We are interested in the mean percent of nitrogen in ancient air and we wish to estimate this quantity.

Plan. We will estimate the mean percent of nitrogen in ancient air by giving a 90% confidence interval.

Solve. It is not clear that these data are an SRS from the late Cretaceous atmosphere, but we are told to assume that they are. The stemplot follows. There are no outliers, and the plot is slightly skewed left. With these few observations, it is difficult to check the assumptions. We might still use the t procedures but perhaps with not as much confidence in their validity as we had in other examples.

```
4 9
5 1 4
5
6 0 3 3 4 4
6 5
```

An approximate 90% confidence interval for the mean percent of nitrogen in ancient air can be calculated from the data on the nine specimens of amber. We use the formula for a t interval, namely $\bar{x} \pm t^* \frac{s}{\sqrt{n}}$. In this problem, $C = 0.90$, $\bar{x} = 59.589$, $s = 6.2553$, $n = 9$; hence t^* is the upper $(1 - 0.90)/2 = 0.05$ critical value for the $t(8)$ distribution. From Table C we see that $t^* = 1.86$. Thus the 90% confidence interval is

$$59.589 \pm 1.86\frac{6.2553}{\sqrt{9}} = 59.589 \pm 3.878 = (55.711, 63.467)$$

Many statistical software packages compute a confidence interval directly, after the data are entered.

Conclude. We are 90% confident that the mean percent of nitrogen in ancient air is between 55.711% and 63.467%.

Exercise 17.28

(a) A 95% confidence interval for the mean systolic blood pressure in the population from which the subjects were recruited can be calculated from the data on the 27 members of the placebo group because they are randomly selected from the 54 subjects. We use the formula for a t interval, namely $\bar{x} \pm t^*\frac{s}{\sqrt{n}}$. In this exercise, $\bar{x} = 114.9$, $s = 9.3$, $n = 27$; hence t^* is the upper $(1 - 0.95)/2 = 0.025$ critical value for the $t(26)$ distribution. From Table C we see $t^* = 2.056$. Thus the 95% confidence interval is

$$114.9 \pm 2.056\frac{9.3}{\sqrt{27}} = 114.9 \pm 3.68 = (111.22, 118.58)$$

(b) For the procedure used in (a), the population from which the subjects were drawn should be such that the distribution of the seated systolic blood pressure in the population is normal. The 27 subjects used for the confidence interval in part (a) should be a random sample from this population. Unfortunately, we do not know if that is the case. Although 27 subjects were selected at random from the total of 54 subjects in the study, we do not know if the 54 subjects were a random sample from this population.

With a sample of 27 subjects, it is not crucial that the population be normal, as long as the distribution is not strongly skewed and the data contain no outliers. It is important that the 27 subjects can be considered a random sample from the population. If not, we cannot appeal to the central limit theorem to ensure that the t procedure is at least approximately correct even if the data are not normal.

(Note: It turns out that since the subjects were divided at random into treatment and control groups, there do exist procedures for comparing the treatment and placebo groups. These procedures are not based on the t distribution, but they are valid as long as treatment groups are determined by randomization. However, the conclusions drawn from these procedures apply only to the subjects in the study. To generalize the conclusions to a larger population, we must know that the subjects are a random sample from this larger population.)

Exercise 17.45

(a) The randomization might be carried out by simply flipping a fair coin. If the coin comes up heads, use the right-hand-threaded knob first. If the coin comes up tails, use the left-hand-threaded knob first. Alternatively, to balance the number of times each type is used first, one might choose an SRS of 12 of the 25 subjects. These 12 use the right-hand-threaded knob first. Everyone else uses the left-hand-threaded knob first.

A second place one might use randomization is in the order in which subjects are tested. Use a table of random digits to determine this order. Label subjects 01 to 25. The first label that appears in the list of random digits (read in groups of two digits) is the first subject measured; the second label that appears is the next subject measured; and so on. This randomization is probably less important than the one described in the previous paragraph. It would be important if the order or time at which a subject was tested might have an effect on the measured response. For example, if the study began early in the morning, the first subject might be sluggish if still sleepy. Sluggishness might lead to longer times and perhaps a larger difference in times. Subjects tested later in the day might be more alert.

(b) *State.* We are interested in whether right-handed people find right-hand threads easier to use than left-hand threads. The experiment actually measures the times in seconds each of 25 right-handed subjects took to move the indicator a fixed distance, once with the left-handed thread and once with the right-handed thread. Presumably shorter times indicate ease of use. Thus, we are interested in whether the times for the left-handed threads are greater than those for the right-handed threads.

Plan. In terms of μ, the mean of the population of differences (left thread time) – (right thread time), we wish to test whether the times for the left-threaded knobs are longer than for the right-threaded knobs;

$H_0: \mu = 0$ and $H_a: \mu > 0$

We will use the one sample significance test based on the t statistic.

Solve. Assuming the randomization we recommended in part (a) is used, this would be a randomized experiment. A histogram of the 25 differences follows. We can see that there are no outliers in the data. The data appear a bit skewed to the left but not strongly enough to threaten the validity of the t procedure given that the sample size is 25. (In the section on the robustness of t procedures, t procedures are safe for samples of size $n \geq 15$ unless there are outliers and/or strong skewness.)

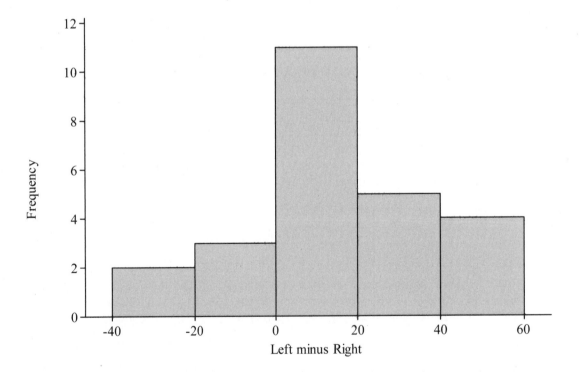

For the 25 differences we compute

$\bar{x} = 13.32$ $\qquad\qquad s = 22.94$

We then use the one sample significance test based on the t statistic.

$$t = \frac{\bar{x} - \mu_0}{s/\sqrt{n}} = \frac{13.32 - 0}{22.94/\sqrt{25}} = 2.903$$

From the value of the t statistic and Table C, the P-value is between 0.0025 and 0.005.

df = 24

p	.005	.0025
t^*	2.797	3.091

Using statistical software, the P-value is computed as P-value = 0.0039.

Conclude. We conclude that there is strong evidence that, on average, the time for left-hand threads is greater than the time for right-hand threads. Assuming that shorter times mean greater ease of use, we would conclude that there is strong evidence that the right-hand threads are easier to use.

Exercise 17.47

$\bar{x}$ = 13.32, s = 22.94, n = 25, and t^* is the upper $(1 - 0.90)/2 = 0.05$ critical value for the $t(24)$ distribution. From Table C, we see that t^* = 1.711. Thus, the 90% confidence interval is

$$13.32 \pm 1.711 \frac{22.94}{\sqrt{25}} = 13.32 \pm 7.85 = (5.47, 21.17)$$

Computing the means, $\bar{x}_R$ = 104.12, $\bar{x}_L$ = 117.44, and $\bar{x}_R / \bar{x}_L$ = 88.7%, so people using the right-handed threads complete the task in about 90% of the time it takes those using the left-handed threads. As an alternative, if for each subject we first take the ratio right-thread/left-thread and then average these ratios, we get 91.7%, which is almost the same answer.

CHAPTER 18

TWO-SAMPLE PROBLEMS

OVERVIEW

One of the most commonly used applications of statistical inference is to compare two means. For example, a researcher may wish to determine which of two methods for teaching children to read is more successful on average. One can compare two means via a confidence interval or via a significance test. The basic ideas for these methods follow those developed in Chapters 14, 15, and 17, but here we extend them to the case in which we collect samples from two populations.

In this setting, for **comparison of two population means,** μ_1 and μ_2, we have two distinct, independent simple random samples from two distinct populations. We select n_1 individuals from the first population, which has standard deviation σ_1, and n_2 observations from the second population, which has standard deviation σ_2. The procedures are based on the difference $\bar{x}_1 - \bar{x}_2$, which is an obvious estimator of the parameter $\mu_1 - \mu_2$. The procedures may be used safely (1) for any total sample size if the two populations are Normal; or (2) with total sample size under (say) 15 if the data appear close to Normal (single peak, roughly symmetric, no outliers); or (3) in any situation with a large (say more than 40) total sample size. About the only situation in which you should absolutely avoid using these procedures is when the total sample size is small (like under 15) and the two populations are heavily skewed. In any case, the procedures described in this chapter are most robust to failures in their assumptions when the sample sizes are equal.

Tests and confidence intervals for the difference in the population means, $\mu_1 - \mu_2$, are based on the **two-sample t statistic.** Despite the name, this test statistic does *not* have an exact t distribution. However there are good approximations to its distribution that allow us to carry out valid significance tests and confidence intervals. Conservative procedures use the $t(k)$ distribution as an approximation where the degrees of freedom k is taken to be the smaller of $n_1 - 1$ and $n_2 - 1$. More accurate procedures use the data to estimate the degrees of freedom k. This procedure is followed by most statistical software.

To carry out a significance test for H_0: $\mu_1 = \mu_2$, use the two-sample t statistic:

$$t = \frac{(\bar{x}_1 - \bar{x}_2)}{\sqrt{\dfrac{s_1^2}{n_1} + \dfrac{s_2^2}{n_2}}}$$

The P-value is found by using the approximate distribution $t(k)$, where k is estimated from the data when using statistical software or can be taken to be the smaller of $n_1 - 1$ and $n_2 - 1$ for a conservative procedure.

An approximate confidence C level **confidence interval** for $\mu_1 - \mu_2$ is given by

$$(\bar{x}_1 - \bar{x}_2) \pm t^* \sqrt{\frac{s_1^2}{n_1} + \frac{s_2^2}{n_2}}$$

where t^* is the upper $(1 - C)/2$ critical value for the $t(k)$ distribution, where k is estimated from the data when using statistical software or can be taken to be the smaller of $n_1 - 1$ and $n_2 - 1$ for a conservative procedure.

The pooled two-sample t procedures are used when we can safely assume that the two populations have equal variances, but this is exceedingly rare. Hence these procedures are generally not recommended. Similarly, the "F-test" for comparing standard deviations of two Normal populations is not recommended because this test is very sensitive to non-Normal distributions. In other words, it's not robust against violations of the assumption of Normal populations.

GUIDED SOLUTIONS

Exercise 18.3

KEY CONCEPTS: Single sample, matched pairs, or two samples

Are there one or two samples involved? Was matching done?

Exercise 18.4

KEY CONCEPTS: Single sample, matched pairs, or two samples

Are there one or two samples involved? Was matching done?

Exercise 18.5

KEY CONCEPTS: Two-sample t significance test; four-step process

We're interested in comparing the number of tree species in unlogged forest areas with that in forest areas logged 8 years ago.

(a) Suppose the companies logging a forest area knew that the impact of logging on the forest environment was being measured. Do you suppose they might behave differently than they do when they don't believe anybody is watching?

(b) We need to carry out a significance test.

The four-step process for tests of significance follows.

State. What is the practical question that requires a statistical test?

Plan. Identify the parameters, state the null and alternative hypotheses, and choose the type of test that fits your situation.

Solve. Carry out the test in three phases:
  (1) Check the conditions for the test you plan to use.
  (2) Calculate the test statistic.
  (3) Find the P-value.

Conclude. Return to the practical question to describe your results in this setting.

To apply the steps to this problem, here are some suggestions:

State. We're comparing two types of forest plots – those that have never been logged, and those that were logged 8 years ago. We're measuring the number of tree species found on these types of forest plots. What is the research question?

Plan. Here we lay out the formal language of the significance testing problem.

First, how many populations are described in this problem? What are they?

What parameters are related to the research question you wrote above?

Remember that the alternative hypothesis is often the conclusion that the researchers suspect. What do you think the researchers suspect?

Write the null and alternative hypotheses for this problem.

What sort of test applies here? Can we assume that the population standard deviations are equal here? Do we have two independent samples?

Solve. First, address the conditions necessary for application of the method described in *Plan* above.

Do we have two independent samples from two distinct populations?

Make a back-to-back stemplot. Are there any clear problems with the assumption that the populations are Normal (or close to it)? Are there any outliers?

Note that in the description of the problem, there's no discussion of how the data were collected. We can't tell whether these samples are SRS's or can be treated as SRS's. If these data were selected in some way that casts doubt on this, then the work we do below can't be trusted. For now, let's assume that these data are close enough to being SRS's from two populations.

Next, compute the summary statistics we need to work the problem.

	Unlogged	Logged
Sample size		
Sample mean		
Sample standard deviation		

Compute the test statistic:

$$T = \frac{\bar{x}_1 - \bar{x}_2}{\sqrt{\dfrac{s_1^2}{n_1} + \dfrac{s_2^2}{n_2}}} =$$

How many degrees of freedom are there? You may use Option 1 or Option 2.

Use Table C to estimate the *P*-value.

Conclude. Can we conclude that on average, logged forest plots have fewer tree species than unlogged forest plots?

Exercise 18.7

KEY CONCEPTS: Two-sample t confidence interval

The formula for a level C confidence interval for $\mu_1 - \mu_2$ is $(\bar{x}_1 - \bar{x}_2) \pm t^* \sqrt{\dfrac{s_1^2}{n_1} + \dfrac{s_2^2}{n_2}}$,

where t^* is the critical value for confidence level C for the t distribution with degrees of freedom equal to the smaller of $n_1 - 1$ and $n_2 - 1$.

In Exercise 18.5, above, you computed the four summary statistics $\bar{x}_1 = 17.5$, $\bar{x}_2 = 13.67$, $s_1 = 3.529$, and $s_2 = 4.5$. We have $n_1 = 12$ and $n_2 = 9$. Using Option 2, you found for 8 degrees of freedom.

What is the critical value, t^*, the critical value for confidence level 90% for the t distribution with 8 degrees of freedom?

$t^* =$

Compute a 90% confidence interval for the mean

$(\bar{x}_1 - \bar{x}_2) \pm t^* \sqrt{\dfrac{s_1^2}{n_1} + \dfrac{s_2^2}{n_2}} =$

Exercise 18.8

KEY CONCEPTS: Two-sample t test

We have two populations: The population of fabric strips buried for two weeks, and the population of fabric strips buried for 16 weeks. Is the mean breaking strength of strips buried 16 weeks smaller than the mean breaking strength of strips buried for 2 weeks? Let μ_1 = mean strength after two weeks. Let μ_2 = mean strength after 16 weeks.

Write the null and alternative hypotheses being tested.

Using Option 2, we have 4 degrees of freedom. The t statistic is 0.988. The P-value is 0.1857.

Does this provide much evidence to conclude that fabric strips buried 16 weeks have lower mean breaking strength than fabric strips buried two weeks?

Exercise 18.14

KEY CONCEPTS: Details for two-sample t test

With respect to Exercise 18.8, above,

Compute the t statistic:

$$t = \frac{\bar{x}_1 - \bar{x}_2}{\sqrt{\dfrac{s_1^2}{n_1} + \dfrac{s_2^2}{n_2}}} =$$

Compute the degrees of freedom using the formula corresponding to Option 1:

$$df = \frac{\left(\dfrac{s_1^2}{n_1} + \dfrac{s_2^2}{n_2}\right)^2}{\dfrac{1}{n_1 - 1}\left(\dfrac{s_1^2}{n_1}\right)^2 + \dfrac{1}{n_2 - 1}\left(\dfrac{s_2^2}{n_2}\right)^2} =$$

Exercise 18.26

KEY CONCEPTS: Two-sample t confidence interval

In some applications, sample standard deviations aren't reported, but standard error of the mean (SEM) is reported. In this problem, sample sizes, means, and SEMs are given. Recall that SEM $= \dfrac{s}{\sqrt{n}}$ so that $s =$ SEM $\sqrt{n}$.

(a)

For the Unrestrained group,

$s_1 = $ SEM$_1 \sqrt{n_1} = $

For the Restrained group,

$s_2 = $ SEM$_2 \sqrt{n_2} = $

(b) Using Option 2, degrees of freedom are the lesser of $n_1 - 1$ and $n_2 - 1$. How many degrees of freedom are there for this procedure?

(c) First determine the critical value t^* for confidence level 90% for a t distribution with the degrees of freedom as given in (b):

$t^* = $

Next, compute the 90% confidence interval:

$$(\bar{x}_1 - \bar{x}_2) \pm t^* \sqrt{\dfrac{s_1^2}{n_1} + \dfrac{s_2^2}{n_2}} = $$

COMPLETE SOLUTIONS

Exercise 18.3

This example involves a single sample. We have a sample of 20 measurements, and we want to see if the mean for this method agrees with the known concentration.

Exercise 18.4

This example involves two samples, the set of measurements on each method. We are not told of any matching.

Exercise 18.5

(a) It is reasonable to guess that if the logging companies knew that the environmental impact of their activities was being observed and measured, they may not behave in the same way that they behave when nobody is watching - perhaps they might take extra care to avoid cutting certain types of trees. We probably can't trust the results if this is the case.

(b)
State. The researchers want to determine whether the mean number of different tree species is lower for forest plots that were logged 8 years ago than for the forest plots that have never been logged.

Plan. There are two populations being studied: the population of all forest plots that have never been logged has mean μ_1, and standard deviation σ_1. The population of all forest plots that were logged 8 years ago has mean μ_2 and standard deviation σ_2. The null and alternative hypotheses are: $H_0 : \mu_1 = \mu_2$, $H_a : \mu_1 > \mu_2$. Since we have two independent samples chosen (ideally randomly) from two distinct populations, the two-sample t significance test seems reasonable, provided the standard conditions necessary for this hold.

Solve. We've discussed the assumption of random samples. Using stemplots we can investigate the assumption that the populations from which we are sampling are not far from Normal, and that there are no outliers. Split stemplots are given:

Logged plots		Unlogged plots
4	0	
	0	
4 2 0	1	3 3
8 8 7 5 5	1	5 5 8 9 9
	2	0 1 2 2
	2	

There's not strong evidence in these plots against the assumption that the populations are close to Normal. Remember that these are small samples, so we have to be somewhat liberal in this assessment. There are no outliers. It seems reasonable to proceed with the two-sample t test.

The summary statistics are:

	Unlogged	Logged
Sample size	12	9
Sample mean	17.5	13.67
Sample standard deviation	3.529	4.5

The test statistic is $t = \dfrac{\bar{x}_1 - \bar{x}_2}{\sqrt{\dfrac{s_1^2}{n_1} + \dfrac{s_2^2}{n_2}}} = \dfrac{17.5 - 13.67}{\sqrt{\dfrac{3.529^2}{12} + \dfrac{4.5^2}{9}}} = 2.112$

The number of degrees of freedom here is the smaller of $12 - 1 = 11$ and $9 - 1 = 8$. Hence, there are 8 degrees of freedom. Looking for two numbers that sandwich 2.112 within the row of Table C corresponding to 8 degrees of freedom, we see that $1.860 < 2.112 < 2.306$. Hence, $0.025 < P\text{-value} < 0.05$.

Conclude. Since our P-value is between .025 and .05, we have pretty strong evidence that the average number of tree species is lower in forest plots logged 8 years ago than in forest plots that have never been logged.

Exercise 18.7

As shown in Exercise 18.5, we have 8 degrees of freedom.

Referring to the data provided with Exercise 18.5, a 90% confidence interval for the difference in mean number of species between unlogged and logged plots is given by;

$$(17.5 - 13.67) \pm 1.860 \sqrt{\frac{3.529^2}{12} + \frac{4.5^2}{9}}$$

or $\qquad 3.83 \pm 3.37$

or $\qquad 0.46$ species to 7.20 species.

Exercise 18.8

Let μ_1 = mean strength after two weeks. Let μ_2 = mean strength after 16 weeks. We test the hypotheses $H_0 : \mu_2 = \mu_1$ and $H_a : \mu_2 < \mu_1$. The t statistic is 0.988. Using Option 2 we have $5 - 1 = 4$ degrees of freedom. Rounding down to be conservative, using Option 1 we also find 4 degrees of freedom. From Table C, t lies between the two critical values 0.941 and 1.190. The P-value is between 0.15 and 0.20. Of course, it's exactly 0.1857, according to the output.

These data provide little evidence to support a conclusion that after 16 weeks buried, fabrics have lower mean breaking strength than that for fabrics buried only two weeks.

Exercise 18.14

Let μ_1 = mean strength after two weeks. Let μ_2 = mean strength after 16 weeks. The two corresponding sample means are $\bar{x}_1$ = 123.8 and $\bar{x}_2$ = 116.4. The two corresponding standard deviations are s_1 = 4.604346 and s_2 = 16.08726.

Entering the summary statistics into the formulas for the t statistic and the approximate degrees of freedom gives:

$$t = \frac{\bar{x}_1 - \bar{x}_2}{\sqrt{\dfrac{s_1^2}{n_1} + \dfrac{s_2^2}{n_2}}} = \frac{123.8 - 116.4}{\sqrt{\dfrac{(4.604346)^2}{5} + \dfrac{(16.08726)^2}{5}}} = 0.9888667$$

$$\text{df} = \frac{\left(\dfrac{s_1^2}{n_1} + \dfrac{s_2^2}{n_2}\right)^2}{\dfrac{1}{n_1 - 1}\left(\dfrac{s_1^2}{n_1}\right)^2 + \dfrac{1}{n_2 - 1}\left(\dfrac{s_2^2}{n_2}\right)^2} = \frac{\left(\dfrac{4.604346^2}{5} + \dfrac{16.08726^2}{5}\right)^2}{\dfrac{1}{4}\left(\dfrac{4.604346^2}{5}\right)^2 + \dfrac{1}{4}\left(\dfrac{16.08726^2}{5}\right)^2} = \frac{3135.999}{674.2685} = 4.650964$$

These values agree very closely with the output provided in Figure 18.5.

Exercise 18.26

(a) Since SEM = $s/\sqrt{n}$, then solving for s we have $s = \text{SEM} \times \sqrt{n}$.

For the Unrestrained group, $s_1 = 7 \times \sqrt{9} = 21$.

For the Restrained group, $s_2 = 10 \times \sqrt{11} = 33.166$.

(b) Under Option 2, the two-sample t procedures conservatively have the lower of 9−1=8 and 11−1=10 degrees of freedom. That is, they have 8 degrees of freedom.

(c) A 90% confidence interval for the mean difference in amount of chips consumed between unrestrained and restrained women is given by:

$$(\bar{x}_1 - \bar{x}_2) \pm t^* \sqrt{\frac{s_1^2}{n_1} + \frac{s_2^2}{n_2}}$$

Using C = .90, the critical value for the t distribution with 8 degrees of freedom is t^* = 1.86. Hence the 90% confidence interval is given by:

$$(59 - 32) \pm 1.860 \sqrt{\frac{21^2}{9} + \frac{33.166^2}{11}} \text{ grams}$$
$$27 \pm 22.7 \text{ grams}$$
$$\text{or } 4.3 \text{ grams to } 49.7 \text{ grams.}$$

With 90% confidence, unrestrained women consume between 4.3 grams and 49.7 grams more potato chips than restrained women.

CHAPTER 19

INFERENCE ABOUT A POPULATION PROPORTION

OVERVIEW

In this chapter, we consider inference about a population proportion p based on the **sample proportion**

$$\hat{p} = \frac{\text{count of successes in the sample}}{\text{count of observations in the sample}}$$

obtained from an SRS of size n, where X is the number of "successes" (occurrences of the event of interest) in the sample. To use the methods of this chapter for inference, the following assumptions need to be satisfied.

• The data are an SRS from the population of interest.
• The population is much larger than the sample.
• The sample size is sufficiently large. Guidelines for sample sizes are given.

In this case, we can treat $\hat{p}$ as having a distribution that is approximately Normal with mean $\mu = p$ and standard deviation $\sigma = \sqrt{p(1-p)/n}$.

An **approximate level C confidence interval** for p is

$$\hat{p} \pm z^* \sqrt{\frac{\hat{p}(1-\hat{p})}{n}}$$

where z^* is the critical value for the standard Normal density curve with area C between $-z^*$ and z^*.

The **standard error** of $\hat{p}$ is given by $\sqrt{\frac{\hat{p}(1-\hat{p})}{n}}$. The **margin of error** associated with the confidence interval described above is $z^* \sqrt{\frac{\hat{p}(1-\hat{p})}{n}}$. Use this interval only when the counts of successes and failures in the sample are both at least 15.

The confidence interval procedure described above is often quite inaccurate unless the sample size is large. A more accurate confidence interval for smaller samples is the **plus four confidence interval.** To get this interval, add four imaginary observations - two successes and two failures - to your sample. Then, with these new values for the number of failures and successes, use the previous formula for the approximate level C confidence interval. Use the plus four confidence interval when the confidence level C is at least 90% and the sample size n is at least 10 (with any combination of successes and failures).

The **sample size** n required to obtain a confidence interval of approximate margin of error m for a proportion is

$$n = \left(\frac{z^*}{m}\right)^2 p^*(1-p^*)$$

where p^* is a guessed value for the population proportion and z^* is the critical value for the standard Normal density curve with area C between $-z^*$ and z^*. To guarantee that the margin of error of the confidence interval is less than or equal to m no matter what the value of the population proportion may be, use a guessed value of $p^* = \frac{1}{2}$.

Tests of the hypothesis $H_0: p = p_0$ are based on the z **statistic**

$$z = \frac{\hat{p} - p_0}{\sqrt{\dfrac{p_0(1-p_0)}{n}}}$$

with P-values calculated from the standard Normal distribution. Use this test when $np_0 \geq 10$ and $n(1 - p_0) \geq 10$.

GUIDED SOLUTIONS

Exercise 19.1

KEY CONCEPTS: Parameters and statistics, proportions

(a) To what group does the study refer?

Population =

Parameter p =

(b) A statistic is a number computed from a sample. What is the size of the sample and how many in the sample said they prayed at least once in a while? From these numbers compute

$$\hat{p} = \frac{\text{count of successes in the sample}}{\text{count of observations in the sample}} =$$

Exercise 19.4

KEY CONCEPTS: When to use the confidence interval procedure for inference about a proportion

Recall the assumptions needed to safely use the methods of this chapter to compute a confidence interval:

- The data are an SRS from the population of interest.

- The population is much larger than the sample.

- For a confidence interval, n is large enough that both the count of successes $n\hat{p}$ and the count of failures $n(1 - \hat{p})$ are 15 or more.

These are the conditions we must check. Are all the conditions met?

Exercise 19.8

KEY CONCEPTS - large sample confidence interval, plus four confidence interval for a proportion

We are interested in estimating with 95% confidence the proportion of American teens with a MySpace profile that post their picture.

(a) The large-sample confidence interval will be given by $\hat{p} \pm z^* \sqrt{\dfrac{\hat{p}(1 - \hat{p})}{n}}$, where $\hat{p}$ is the sample proportion of successes, n is the sample size, and z^* is the critical value for the standard Normal density curve with area .9500 between $-z^*$ and z^*. In this problem,

$$n =$$

$$\hat{p} =$$

and for 95% confidence, using Table C,

$$z^* =$$

Use these to construct a 95% large-sample confidence interval for the proportion p of all teens with profiles who include photos of themselves:

(b) The "plus four" estimate of p is given by $\tilde{p} = \dfrac{\text{number of successes in the sample } +2}{n+4}$. The plus-four

95% confidence interval for p is given by $\tilde{p} \pm z^* \sqrt{\dfrac{\tilde{p}(1-\tilde{p})}{n+4}}$, where n and z^* are unchanged from part (a).

First, compute the estimate: $\tilde{p} =$

Construct the plus four 95% confidence interval for p:

Finally, compare the two confidence intervals for p you constructed. Are the margins of error almost the same? What is the difference between these confidence intervals?

Exercise 19.11

KEY CONCEPTS: Sample size, margin of error

The sample size n required to obtain a confidence interval of approximate margin of error m for a proportion is

$$n = \left(\frac{z^*}{m}\right)^2 p^*(1-p^*)$$

where p^* is a guessed value for the population proportion and z^* is the critical value of the standard Normal distribution for the desired level of confidence. To apply this formula here we must determine

m = desired margin of error =

p^* = a guessed value for the population proportion =

z^* = critical value needed for a 90% confidence interval =

From the statement of the exercise, what are these values? Once you have determined them, use the formula to compute the required sample size n.

$$n = \left(\frac{z^*}{m}\right)^2 p^*(1-p^*) =$$

Exercise 19.14

KEY CONCEPTS: When to use the z test for a proportion

Recall that the (large sample) z test for a proportion is appropriate if
    (i) the sample can be considered an SRS
    (ii) the population we're sampling from is much larger than the sample
    (iii) both $np_0 \geq 10$ and $n(1 - p_0) \geq 10$.
These are the conditions we must check in (a) and (b).

(a)

(b)

Exercise 19.40

KEY CONCEPTS: Confidence interval for a proportion; the four-step process.

The four step process follows.

    State. What is the practical question that requires estimating a parameter?

    Plan. Identify the parameter, choose a level of confidence, and select the type of confidence interval that fits your situation.

    Solve. Carry out the test in two phases:
        1. Check the conditions for the interval you plan to use.
        2. Calculate the confidence interval.

    Conclude. Return to the practical question to describe your results in this setting.

To apply the steps to this problem, here are some suggestions. You can use Example 19.5 in the text as a guide.

State. What is the population being studied in this problem?

    What do the researchers hope to estimate?

Plan. What parameter are the researchers interested in estimating?

What is the level of confidence to be used here?

In this section, two different confidence interval forms were considered. Which one is recommended? Write the formula here:

Solve. First, check conditions:

Can we consider the sample to be a SRS from the population?

Is the population much larger than the sample?

Is the sample large enough? Your answer here will depend upon the confidence interval method you chose in *Plan*, above.

Our sample size is $n = 117$. Of these, 68 use a seatbelt.
Compute the estimate corresponding to the confidence interval method selected in *Plan*, above.

What is the critical value needed for a 95% confidence interval?

$z^* =$

Compute the appropriate 95% confidence interval:

Conclude. State any conclusions in the context of this problem.

Exercise 19.42

KEY CONCEPTS: Testing hypotheses about a proportion; the four-step process.

The four step process for tests of significance follows.

State. What is the practical question that requires a statistical test?

Plan. Identify the parameter, state null and alternative hypotheses, and choose the type of test that fits your situation.

Solve. Carry out the test in three phases:
(1) Check the conditions for the test you plan to use,
(2) Calculate the test statistic,
(3) Find the *P*-value.

Conclude. Return to the practical question to describe your results in this setting.

To apply the steps to this problem, here are some suggestions. Use Example 19.7 in the text as a guide.

State. State the problem.

Plan. What is the parameter of interest?

What does the researcher suspect, or what is he/she trying to show?

Write the null and alternative hypotheses of interest:

What type of test should be used?

Solve. First check that the appropriate conditions for inference are satisfied.

Is the sample an SRS, or can it be treated as such?

Is the sample size much smaller than the population size?

Are both $np_0 \geq 10$ and $n(1-p_0) \geq 10$?

Compute the sample proportion of female Hispanic drivers in Boston who wear seatbelts.

$$\hat{p} =$$

Calculate the test statistic:

$$z = \frac{\hat{p} - p_0}{\sqrt{\dfrac{p_0(1 - p_0)}{n}}} =$$

Find the *P*-value:

$$P\text{-value} =$$

Conclude. In the context of this problem, what do you conclude?

COMPLETE SOLUTIONS

Exercise 19.1

(a) The population is presumably all college students. The parameter p is the proportion of all college students who pray at least once in a while.

(b) The statistic is $\hat{p}$, the proportion in the sample who said that they prayed at least once in a while.

$$\hat{p} = 107/127 = 0.8425$$

Exercise 19.4

Though it is not explicitly stated, there may be little reason to believe that we can't treat this sample as an SRS from the population of interest. The population of adult heterosexuals is extremely large compared with the sample size of 2673 adult heterosexuals. However, the number of successes is $n\hat{p} = 2673 \times 0.002 = 5.346$. We don't have at least 15 successes in the sample. We can't use the large-sample confidence interval to estimate the proportion p who share these two risk factors.

Exercise 19.8

(a) The sample proportion of successes is $\hat{p} = \dfrac{385}{487} = 0.7906$.   Using Table C, the critical value needed for a 95% confidence interval is $z^* = 1.96$.   Hence, a 95% large-sample confidence interval for the proportion of teens with MySpace profiles that posted photos of themselves is given by

$$\hat{p} \pm z^* \sqrt{\dfrac{\hat{p}(1-\hat{p})}{n}}$$

$$0.7906 \pm 1.96 \sqrt{\dfrac{0.7906 \times (1-0.7906)}{487}}$$

or 0.7906 ± 0.0361
or 0.7545 to 0.8267

(b) The plus four estimate of p is $\tilde{p} = \dfrac{\text{number of successes in the sample} + 2}{n+4} = \dfrac{385+2}{487+4} = 0.7882$.
The corresponding plus four confidence interval for p is

$$\tilde{p} \pm z^* \sqrt{\dfrac{\tilde{p}(1-\tilde{p})}{n+4}}$$

$$0.7882 \pm 1.96 \sqrt{\dfrac{0.7882 \times (1-0.7882)}{487+4}}$$

or 0.7882 ± 0.0361
or 0.7521 to 0.8243

The margins of error with these intervals agree to at least four decimal places. The plus four estimate pulls the ordinary sample proportion toward 0.50, so the interval in (b) is shifted slightly.

Exercise 19.11

We start with the guess that $p^* = 0.75$. For 90% confidence we use $z^* = 1.645$. The sample size we need for a margin of error $m = 0.04$ is thus

$$n = \left(\dfrac{z^*}{m}\right)^2 p^*(1-p^*) = \left(\dfrac{1.645}{0.04}\right)^2 0.75(1-0.75) = 317.11$$

We round up to $n = 318$. Thus, a sample of size 318 is needed to estimate the proportion of Americans with at least one Italian grandparent who can taste PTC to within ± 0.04 with 90% confidence.

Exercise 19.14

(a) We see that $np_0 = (10)(0.5) = 5 < 10$, so the normal approximation to the binomial should *not* be used in this case.

(b) We see that $np_0 = (200)(0.99) = 198 \geq 10$ and $n(1-p_0) = (200)(1-0.99) = (200)(0.01) = 2 < 10$. The normal approximation to the binomial should *not* be used in this case.

Exercise 19.40

State. Of all Hispanic female drivers in Boston, what proportion use seatbelts?

Plan. Let p denote the unknown proportion of all Hispanic female drivers in Boston who use seatbelts. We will construct a 95% confidence interval for this proportion. We should use the plus four confidence interval $\tilde{p} \pm z^* \sqrt{\dfrac{\tilde{p}(1-\tilde{p})}{n+4}}$, where $\tilde{p} = \dfrac{\text{number of successes in the sample} + 2}{n+4}$. This is a more accurate confidence interval than the more traditional large-sample confidence interval also described in the text.

Solve. First, we check whether conditions necessary for use of this method are met. Depending on how the 117 Hispanic female drivers in our sample were chosen, it might be reasonable to treat this as a SRS of all Hispanic female drivers in Boston. (It's easy, however, to believe that this is not a SRS: Suppose, for example, that the sample consists only of motorists that were pulled over by a police officer for a moving violation. It's doubtful that violating motorists represent all motorists.) We will use the 95% confidence level, which is larger than the required 90% confidence level. Finally, our sample of 117 Hispanic female drivers in Boston is larger than the required 10. All conditions are satisfied.

For our sample, $\tilde{p} = \dfrac{68+2}{117+4} = 0.5785$. The required critical value is $z^* = 1.96$. Hence, a 95% confidence interval for the proportion p of Hispanic female drivers in Boston that use seatbelts is

$$\tilde{p} \pm z^* \sqrt{\frac{\tilde{p}(1-\tilde{p})}{n+4}}$$
$$0.5785 \pm 1.96 \sqrt{\frac{0.5785 \times (1-0.5785)}{117+4}}$$
$$0.5785 \pm 0.0880$$

$$\text{or} \quad 0.491 \text{ to } 0.667.$$

Conclude. We estimate with 95% confidence that between about 49% and 67% of all Hispanic female drivers in Boston use seatbelts.

Exercise 19.42

State. We would like to know if more than 50% of Hispanic female drivers in Boston wear seatbelts.

Plan. Let p be the proportion of all Hispanic female drivers in Boston who wear seatbelts. The researcher wonders whether this proportion is larger than 0.5. We want to test the hypotheses

$$H_0: p = 0.5 \qquad H_a: p > 0.5$$

We'll use the large-sample significance test (z test) for a proportion.

Solve. The sample is assumed to be a random sample of all Hispanic female drivers in Boston. The sample of $n = 117$ drivers is reasonably large, but is obviously much smaller than the population of all Hispanic female drivers in Boston. Now, $np_0 = (117)(0.5) = 58.5 \geq 10$ and $n(1 - p_0) = 117(1 - 0.5) = 58.5 \geq 10$, so the conditions for inference are met.

Investigators observed a random sample of 117 Hispanic female drivers and found that 68 of these drivers were wearing seatbelts. In our sample, the proportion of Hispanic female drivers wearing seatbelts was

$$\hat{p} = 68/117 = 0.5812$$

The computed test statistic is

$$z = \frac{\hat{p} - p_0}{\sqrt{\dfrac{p_0(1 - p_0)}{n}}} = \frac{0.5812 - 05}{\sqrt{\dfrac{0.5(1 - 0.5)}{117}}} = \frac{0.0812}{0.0462} = 1.76$$

The *P*-value is the area under the standard Normal density to the right of $z = 1.76$, which is $1 - 0.9608 = 0.0392$.

Conclude. There is reasonably strong evidence that more than half of all Hispanic female drivers in Boston wear seatbelts.

Note: You might wonder why in Problem 19.40 above, "50%" was contained in a 95% confidence interval for *p*, while in Problem 19.42 we reject "50%" as a plausible value for *p* at the 5% level of significance. The most important reason is that inferences made from confidence intervals such as the one used in Problem 19.40 above coincide with two-sided tests of significance. Indeed, if in Problem 19.42 we had a two-sided alternative hypotheses H_a: $\mu \neq .50$, the corresponding *P*-value would have been about .078, and we would not have rejected $p = .50$ as plausible.

CHAPTER 20

COMPARING TWO PROPORTIONS

OVERVIEW

Confidence intervals and tests designed to compare two population proportions are based on the **difference in the sample proportions** $\hat{p}_1 - \hat{p}_2$. The formula for the level C confidence interval is

$$\hat{p}_1 - \hat{p}_2 \pm z^* \, SE$$

where z^* is the critical value for the standard Normal density with area C between $-z^*$ and z^*, and SE is the standard error for the difference in the two proportions computed as

$$SE = \sqrt{\frac{\hat{p}_1(1-\hat{p}_1)}{n_1} + \frac{\hat{p}_2(1-\hat{p}_2)}{n_2}}$$

In practice, use this confidence interval when the populations are at least 10 successes and at least 10 failures in both samples, both of which are simple random samples from large populations.

To get a more accurate confidence interval, especially for smaller samples, add four imaginary observations - one success and one failure - in each sample. Then, with these new values for the number of failures and successes, use the previous formula for the approximate level C confidence interval. This is the **plus four confidence interval.** You can use it whenever both samples have five or more observations.

Significance tests for the equality of the two proportions, $H_0: p_1 = p_2$, use a different standard error for the difference in the sample proportions, which is based on a **pooled estimate** of the common (under H_0) value of p_1 and p_2,

$$\hat{p} = \frac{\text{count of successes in both samples combined}}{\text{count of observations in both samples combined}}$$

The test uses the z statistic

$$z = \frac{\hat{p}_1 - \hat{p}_2}{\sqrt{\hat{p}(1-\hat{p})\left(\dfrac{1}{n_1} + \dfrac{1}{n_2}\right)}},$$

and *P*-values are computed using Table A of the standard normal distribution. In practice, use this test when the populations are at least 10 times as large as the samples and the counts of successes and failures are five or more in both samples.

GUIDED SOLUTIONS

Exercise 20.21

KEY CONCEPTS: Testing equality of two population proportions

First verify that it is safe to use the *z* test for equality of two proportions.

Let p_1 represent the proportion of papers *without* statistical assistance that were rejected without being reviewed in detail, and p_2 the proportion of papers *with* statistical help that were rejected without being reviewed in detail.   Recall that a test of the hypothesis $H_0: p_1 = p_2$ uses the *z* statistic

$$z = \frac{\hat{p}_1 - \hat{p}_2}{\sqrt{\hat{p}(1-\hat{p})\left(\dfrac{1}{n_1} + \dfrac{1}{n_2}\right)}}$$

where n_1 and n_2 are the sizes of the samples, $\hat{p}_1$ and $\hat{p}_2$ are the estimates of p_1 and p_2, and

$$\hat{p} = \frac{\text{count of successes in both samples combined}}{\text{count of observations in both samples combined}}$$

First state the hypotheses to be tested. Is the alternative hypothesis one-sided or two-sided?

The two sample sizes are

$n_1 =$

$n_2 =$

From the data, the estimates of these two proportions are

$\hat{p}_1 =$

$\hat{p}_2 =$

Compute the pooled estimate of the value common to p_1 and p_2 under H_0:

$$\hat{p} = \frac{\text{count of successes in both samples combined}}{\text{count of observations in both samples combined}} =$$

Compute the test statistic:

$$z = \frac{\hat{p}_1 - \hat{p}_2}{\sqrt{\hat{p}(1-\hat{p})\left(\dfrac{1}{n_1} + \dfrac{1}{n_2}\right)}} =$$

Compute the P-value:

P-value =

What do you conclude?

Exercise 20.23

KEY CONCEPTS: Large sample confidence interval for the difference of two population proportions

First determine whether the conditions for the large sample confidence interval are met or whether the plus four confidence interval needs to be used.

The two populations are proportions of papers rejected without review when a statistician *is* and *is not* involved in the research. The two sample sizes are

n_1 = number of papers rejected without review *without* a statistician involved =

n_2 = number of papers rejected without review *with* a statistician involved =

and the number of "successes" are

Number of papers in sample rejected without review *without* a statistician involved =

Number of papers in sample rejected without review *with* a statistician involved =

From the data, the estimates of the two proportions are

$\hat{p}_1 =$

$\hat{p}_2 =$

Let p_1 represent the proportion of all papers rejected without review *without* a statistician involved, and p_2 represent the proportion of all papers rejected without review *with* a statistician involved. Recall that a level C confidence interval for $p_1 - p_2$ is

$$(\hat{p}_1 - \hat{p}_2) \pm z^*SE$$

where z^* is the appropriate critical value for the standard Normal density, and SE is the standard error for the difference in the two proportions computed as

$$SE = \sqrt{\frac{\hat{p}_1(1-\hat{p}_1)}{n_1} + \frac{\hat{p}_2(1-\hat{p}_2)}{n_2}}$$

Use the values of $\hat{p}_1$ and $\hat{p}_2$ you computed to obtain the standard error:

$$SE = \sqrt{\frac{\hat{p}_1(1-\hat{p}_1)}{n_1} + \frac{\hat{p}_2(1-\hat{p}_2)}{n_2}} =$$

For a 95% confidence interval,

$z^* =$

Compute the interval:

$(\hat{p}_1 - \hat{p}_2) \pm z^*SE$:

Exercise 20.27

KEY CONCEPTS: Testing equality of two population proportions; the four-step process

The four step process for testing hypotheses follows.

State. What is the practical question that requires a statistical test?

Plan. Identify the parameters, state null and alternative hypotheses, and choose the type of test that fits your situation.

Solve. Carry out the test in three phases:
(1) Check the conditions for the test you plan to use,
(2) Calculate the test statistic,
(3) Find the *P*-value.

Conclude. Return to the practical question to describe your results in this setting.

To apply the steps to this problem, here are some suggestions. You may want to use Examples 20.4 and 20.5 of the text as a guide.

State. Describe the problem of interest and the data obtained.

Plan. Are there two populations being compared in this problem? What are they?

Define the two proportions of interest.

Is the alternative hypothesis one-sided or two-sided?

Write the null and alternative hypotheses.

What kind of test will you use?

Solve. First check the conditions for using the test.

Write the two sample sizes.

$n_1 =$

$n_2 =$

From the data, the estimates of the two proportions are

$\hat{p}_1 =$

$\hat{p}_2 =$

Compute the pooled estimate of the value common to p_1 and p_2 under H_0:

$$\hat{p} = \frac{\text{count of successes in both samples combined}}{\text{count of observations in both samples combined}} =$$

Now compute the test statistic:

$$z = \frac{\hat{p}_1 - \hat{p}_2}{\sqrt{\hat{p}(1-\hat{p})\left(\dfrac{1}{n_1} + \dfrac{1}{n_2}\right)}} =$$

Compute the *P*-value:

P-value $=$

Conclude. What do you conclude?

Exercise 20.33

KEY CONCEPTS: Plus four confidence intervals for the difference between two population proportions

The four-step process for a confidence interval follows.

State. What is the practical question that requires estimating a parameter?

Plan. Identify the parameters, choose a level of confidence, and select the type of confidence interval that fits your situation.

Solve. Carry out the work in two phases:
    (1) Check the conditions for the interval you plan to use.
    (2) Calculate the confidence interval.

Conclude. Return to the practical question to describe our results in this setting.

To apply the steps to this problem, here are some suggestions. You may want to use Example 20.3 of the text as a guide.

State. What are the populations of interest? What is it we want to estimate?

Plan. What are the parameters of interest in this problem?

What are we going to estimate using a confidence interval? Write this in terms of the parameters.

What confidence interval will you use? Remember that the Plus Four confidence interval is recommended over the traditional large-sample confidence interval. Write the formula for this interval:

Solve.

Are conditions required for use of the confidence interval method you selected in "Plan" (above) satisfied?

Compute the two proportion estimates.

$\hat{p}_1 =$

$\hat{p}_2 =$

Compute the standard error.

$SE =$

For a 90% confidence interval, $z* = 1.645$, so our 90% confidence interval is

$(\hat{p}_1 - \hat{p}_2) \pm z*SE$:

Conclude. What can you conclude about the difference in proportions between the two areas?

COMPLETE SOLUTIONS

Exercise 20.21

The count of successes and failures are each five or more in both samples, so the z test for equality of two population proportions can be used.

We are interested in determining whether there is good evidence that the proportion of papers rejected without review is *different* for papers with and without statistical help. Thus, the hypotheses to be tested are

$$H_0: p_1 = p_2$$

$$H_a: p_1 \neq p_2$$

Letting Population 1 be papers *without* statistical help and Population 2 be papers *with* statistical help, the two sample sizes and estimates of the proportions are

$$n_1 = 190 \qquad \hat{p}_1 = 135/190 = 0.7105$$

$$n_2 = 514 \qquad \hat{p}_2 = 293/514 = 0.5700$$

The pooled sample proportion is

$$\hat{p} = \frac{\text{count of successes in both samples combined}}{\text{count of observations in both samples combined}} = \frac{135 + 293}{190 + 514} = \frac{428}{704} = 0.6080$$

and the z statistic is

$$z = \frac{\hat{p}_1 - \hat{p}_2}{\sqrt{\hat{p}(1-\hat{p})\left(\dfrac{1}{n_1} + \dfrac{1}{n_2}\right)}} = \frac{0.7105 - 0.5700}{\sqrt{0.6080(1 - 0.6080)\left(\dfrac{1}{190} + \dfrac{1}{514}\right)}} = \frac{0.1405}{0.0414} = 3.39$$

Finally, using Table A, we have

$$P\text{-value} = 2 \times P(z \geq 3.39) = (2)(1 - 0.9997) = 0.0006$$

which is very strong evidence that a higher proportion of papers submitted without statistical help will be sent back without review. However, because this is an observational study, the evidence does not establish causation - getting statistical help for your paper may not make it less likely to be sent back without review.

Exercise 20.23

The counts of successes and failures are each 10 or more in both samples, so we may use the large-sample confidence interval.

Let Population 1 be papers *without* statistical help, and Population 2 be papers *with* statistical help. Let a success denote the event that a paper is sent back without review. Then for each population, sample sizes, number of successes and estimates of the proportions follow.

Population 1 $n_1 = 190$ number of successes = 135 $\hat{p}_1 = 135/190 = 0.7105$

Population 2 $n_2 = 514$ number of successes = 293 $\hat{p}_2 = 293/514 = 0.5700$

Using these values, the standard error is

$$SE = \sqrt{\frac{\hat{p}_1(1-\hat{p}_1)}{n_1} + \frac{\hat{p}_2(1-\hat{p}_2)}{n_2}} = \sqrt{\frac{0.7105(1-0.7105)}{190} + \frac{0.5700(1-0.5700)}{514}} = \sqrt{0.001559} = 0.0395$$

For a 95% confidence interval, $z^* = 1.96$. The 95% confidence interval is

$$(\hat{p}_1 - \hat{p}_2) \pm z^*SE = (0.7105 - 0.5700) \pm (1.96)(0.0395) = 0.1405 \pm 0.0774 = 0.0631 \text{ to } 0.2179$$

We are 95% confident that the percentage of papers without statistical help that are not reviewed is between 6.3 and 21.8 percentage points higher than the percentage of papers with statistical help.

Exercise 20.27

State. North Carolina University looked at factors that affected the success of students in a required chemical engineering course. Students must receive a C or better in the course to continue as chemical engineering majors, so we consider a grade of C or better as a success. Is there a difference in the proportions of male and female students who succeeded in the course? The data showed that 23 of the 34 women and 60 of the 89 men succeeded. We view these as SRSs of men and women who would take this course.

Plan. Let p_1 denote the proportion of female students who will succeed, and p_2 the proportion of males who will succeed. We are interested in determining whether there is a difference in these two proportions; hence we test the hypotheses

$$H_0: p_1 = p_2$$
$$H_a: p_1 \neq p_2$$

Solve. We can use the z test when the count of successes and failures are each 5 or more in both samples. For the females there were 23 successes and $34 - 23 = 11$ failures, and for the males there were 60 successes and $89 - 60 = 29$ failures. The conditions for safely using the test are met.

The two sample sizes are

n_1 = number of female students in the course = 34

n_2 = number of male students in the course = 89

From the data, the estimates of these two proportions are

$\hat{p}_1 = 23/34 = 0.6765$

$\hat{p}_2 = 60/89 = 0.6742$

We compute

$$\hat{p} = \frac{\text{count of successes in both samples combined}}{\text{count of observations in both samples combined}} = \frac{23+60}{34+89} = 83/123 = 0.6748$$

The value of the z-test statistic is thus

$$z = \frac{\hat{p}_1 - \hat{p}_2}{\sqrt{\hat{p}(1-\hat{p})\left(\dfrac{1}{n_1}+\dfrac{1}{n_2}\right)}} = \frac{0.6765 - 0.6742}{\sqrt{0.6748(1-0.6748)\left(\dfrac{1}{34}+\dfrac{1}{89}\right)}} = \frac{0.0023}{\sqrt{0.00892}} = 0.02$$

We compute the P-value using Table A (we need to double the tail area because this is a two-sided test):

P-value $= 2 \times (0.4920) = 0.9840$

Conclude. The data provide no evidence of a difference between the proportion of men and women who succeed.

Exercise 20.33

State. What is the size of the difference in the proportion of mice ready to breed in good acorn years and bad acorn years?

Plan. In a low-acorn year, experimenters added hundreds of thousands of acorns to Area 1 to simulate a good acorn year, while Area 2 was left untouched. They then trapped mice in both areas and counted the number of mice in breeding condition. The data follow.

Population	Population description	Sample size	Number of successes	Sample proportion
1	Area 1 (acorn rich)	$n_1 = 72$	54	$\hat{p}_1 = 54/72 = 0.75$
2	Area 2 (acorn poor)	$n_2 = 17$	10	$\hat{p}_2 = 10/17 = 0.59$

We want a 90% confidence interval for the difference of population proportions $p_1 - p_2$, where p_1 is the proportion of mice ready to breed in good acorn years, and p_2 is the proportion of mice ready to breed in bad acorn years. We'll use the plus four confidence interval procedure.

Solve. We can use the large-sample confidence interval when the populations are much larger than the samples and the counts of successes and failures are 10 or more in both samples. That isn't the case here. Anyway, the plus four confidence interval procedure is more accurate, and it is appropriate when each sample has size at least 5. That procedure is appropriate here. To apply the plus four confidence interval procedure, we add four imaginary observations. The new data summary follows.

Population	Population description	Sample size	Number of successes	Plus four sample proportion
1	Area 1 (acorn rich)	$n_1 + 2 = 74$	$54 + 1 = 55$	$\hat{p}_1 = 55/74 = 0.74$
2	Area 2 (acorn poor)	$n_2 + 2 = 19$	$10 + 1 = 11$	$\hat{p}_2 = 11/19 = 0.58$

The standard error is

$$SE = \sqrt{\frac{\hat{p}_1(1-\hat{p}_1)}{n_1} + \frac{\hat{p}_2(1-\hat{p}_2)}{n_2}} = \sqrt{\frac{0.74(1-0.74)}{74} + \frac{0.58(1-0.58)}{19}} = \sqrt{0.0026 + 0.0128} = 0.12$$

and for a 90% confidence interval $z^* = 1.645$, so the 90% confidence interval is

$$(\hat{p}_1 - \hat{p}_2) \pm z^*\text{SE} = (0.74 - 0.58) \pm 1.645(0.12) = 0.16 \pm 0.20 \text{ or } -0.04 \text{ to } 0.36.$$

Conclude. We are 90% confident that mice in Area 1 (with abundant crop) are between –4% and 36% more likely to be in breeding condition than those in Area 2, which was left untouched. The confidence interval is quite wide and there isn't much evidence of a difference in the breeding condition of mice in the two areas in either direction.

CHAPTER 21

INFERENCE ABOUT VARIABLES: PART III REVIEW

To assist you in reviewing the material in Chapters 17–20, we provide the text chapter and related problems in this Study Guide for each of the odd-numbered review exercises. Other than pointing you in the right direction, we provide no additional hints or solutions. At this point, you should be able to work these problems on your own with minimal assistance. As a final challenge, we encourage you to work some of the Supplementary Exercises, which integrate more fully the material in these chapters.

Exercise 21.1
Text Location – Chapter 19 for one-sample confidence interval for p
Related Study Guide exercises – Exercises 19.8, 19.40

Exercise 21.3
Text Location – Chapter 19 for one-sample significance test for p.
Related Study Guide exercises – Exercise 19.42

Exercise 21.5
Text Location – Chapter 17 for one-sample t confidence interval for a mean
Related Study Guide exercises – Exercises 17.7, 17.28

Exercise 21.7
Text Location – Chapter 18 for two-sample t significance test for two means
Related Study Guide exercises – Exercise 18.5

Exercise 21.9
Text Location – Chapter 20 for two-sample significance test for two proportions
Related Study Guide exercises – Exercises 20.21, 20.27

Exercise 21.11
(a) Text Location – Chapter 18 for robustness of two-sample t-distribution methods
    Related Study Guide exercises – Exercise 18.5
(b) Text Location – Chapter 18 for two-sample t significance test for two means
    Related Study Guide exercises – Exercises 18.5, 18.8, 18.14

Exercise 21.13
Text Location – Chapter 18 for two-sample t confidence interval for a difference of means
Related Study Guide exercises – Exercises 18.7, 18.26

Exercise 21.15
Text Location – Chapter 19 for one-sample confidence interval for *p*
Related Study Guide exercises – Exercises 19.8, 19.40

Exercise 21.17
Text Location – Chapter 17 for one-sample *t* significance test for a mean
Related Study Guide exercises – Exercise 17.7, 17.28

Exercise 21.19
(a) Text Location – Chapter 9 for observational studies and experiments
     Related Study Guide exercises – Exercises 9.1, 9.12
(b) Text Location – Chapter 20 for two-sample significance test for two proportions
     Related Study Guide exercises – Exercises 20.21, 20.27

Exercise 21.21
Text Location – Chapter 20 for two-sample significance test for two proportions; Chapter 18 for two-sample *t* significance test for two means
Related Study Guide exercises – Exercises 20.21, 20.27, 18.5, 18.8, 18.14

Exercise 21.23
Text Location – Chapter 20 for two-sample significance test for two proportions
Related Study Guide exercises – Exercises 20.21, 20.27

Exercise 21.25
Text Location – Chapter 18 for two-sample *t* significance test for two means
Related Study Guide exercises – Exercises 18.5, 18.8, 18.14

Exercise 21.27
Text Location – Chapter 17 for one-sample *t* confidence interval
Related Study Guide exercises – Exercises 17.7, 17.28

Exercises 21.29, **21.31**, **21.33**, and **21.35** are designed to assess your ability to determine when various statistical methods discussed in Chapters 17–20 should be applied. We won't point you to problems from these chapters. Instead, we suggest that you should read the Overview to these chapters in the Study Guide.

CHAPTER 22

TWO CATEGORICAL VARIABLES: THE CHI-SQUARE TEST

OVERVIEW

The inference methods of Chapter 20 of your text are extended to a comparison of more than two population proportions. When comparing more than two proportions, it is best to first do an overall test to see if there is good evidence of any differences among the proportions. Then a detailed follow-up analysis can be performed to decide which proportions are different and to estimate the sizes of the differences.

The overall test for comparing several population proportions arranges the data in a **two-way table.** Two-way tables were introduced in Chapter 6 of your text; the tables are a way of displaying the relationship between any two categorical variables. The tables are also called **$r \times c$ tables,** where **r** is the number of rows and **c** is the number of columns. Often the rows of the two-way tables correspond to populations or treatment groups, the columns to different categories of the response. When comparing several proportions, there would be only two columns since the response takes only one of two values, and the two columns would represent the successes and failures.

The null hypothesis is $H_0 : p_1 = p_2 = \cdots = p_n$, which says that the n population proportions are the same. The alternative is "many sided," as the proportions can differ from each other in a variety of ways. To test this hypothesis, we will compare the **observed counts** with the **expected counts** when H_0 is true. The expected cell counts are computed using the formula

$$\text{expected count} = \frac{\text{row total} \times \text{column total}}{n}$$

where n is the total number of observations.

The statistic we will use to compare the expected counts with the observed counts is the **chi-square statistic.** It measures how far the observed and expected counts are from each other using the formula

$$X^2 = \sum \frac{(\text{observed} - \text{expected})^2}{\text{expected}}$$

where we sum up all the $r \times c$ cells in the table.

When the null hypothesis is true, the distribution of the test statistic X^2 is approximately the chi-squared distribution with $(r-1)(c-1)$ degrees of freedom. The P-value is the area to the right of X^2 under the chi-square density curve. Use Table D in the Appendix of the book to get the critical values and to compute the P-value. The mean of any chi-square distribution is equal to its degrees of freedom.

We can use the chi-square statistic when the data satisfy the following conditions:

- The data are independent SRSs from several populations and each observation is classified according to one categorical variable.

- The data are from a single SRS and each observation is classified according to two categorical variables.

- No more than 20% of the cells in the two-way table have expected counts less than 5.

- All cells have an expected count of at least 1.

- In the special case of the 2 × 2 table, all expected counts should exceed 5 before applying the approximation.

GUIDED SOLUTIONS

Exercise 22.1

KEY CONCEPTS: Conditional distributions in two-way tables

(a) The data from Exercise 22.1 are reproduced here to help you carry out the calculations. We have included a Total row that gives the column totals, which are required to obtain the conditional distributions.

Campus

Facebook category	University Park	Commonwealth
Do not use Facebook	68	248
Several times a month or less	55	76
At least once a week	215	157
At least once a day	640	394
Total	978	875

You are asked to the percent of University Park students that fall in each Facebook category. There are a total of 978 students on the University Park campus, and of these, 68 do not use Facebook. Thus, the percent of University Park students who "Do not use Facebook" is 68/978 = 0.070. Now, compute the percent of University Park students in the other three Facebook categories. The total of the percents should add to 100%, except for possible roundoff error. Next, find the percent of Commonwealth students who fall in each category.

Campus

Facebook category	University Park	Commonwealth
Do not use Facebook	0.070	
Several times a month or less		
At least once a week		
At least once a day		
Total		

(b) In the space below, sketch the conditional distributions in a bar graph similar to Figure 22.1, or use your software to produce the bar graph.

What are the most important differences in Facebook use between the two campus settings?

Exercise 22.5

KEY CONCEPTS: Computing and interpreting expected counts in a two-way table

(a) The data from Exercise 22.1 are reproduced here to help you carry out the calculations. The category "Never use Facebook" has been eliminated. We have included a Total column that gives the row totals and a Total row that gives the column totals, which are required to obtain the expected counts.

Campus

Facebook category	University Park	Commonwealth	Total
Several times a month or less	55	76	131
At least once a week	215	157	372
At least once a day	640	394	1034
Total	910	627	1537

Remember that the expected cell counts are computed using the formula

$$\text{expected count} = \frac{\text{row total} \times \text{column total}}{n}$$

where n is the total number of observations. The value in the Commonwealth column for "Several times a month or less" was obtained as

$$\text{expected count} = \frac{\text{row total} \times \text{column total}}{n} = \frac{(131)(627)}{1537} = 53.44$$

Facebook category	Commonwealth
Several times a month or less	53.44
At least once a week	
At least once a day	

Fill in the rest of the expected counts. Verify that the column total of expected counts agrees with the row total of observed counts.

(b) Which cells have the largest deviations between expected and observed counts? What kind of relationship do these suggest?

Exercise 22.7

KEY CONCEPTS: Computing the X^2 statistic

(a) What is the cell count requirement for use of chi-square? Is it satisfied for these data?

(b) What hypotheses does the chi-square test in this example?

Give the test statistic and P-value from the Minitab output. What do you conclude?

(c) The terms contributing most to X^2 are in the monthly and daily row. What are these telling you about the most important differences in Facebook use between students at the two locations?

Exercise 22.11

KEY CONCEPTS: Degrees of freedom for the chi-square distribution, P-values using Table D

(a) What are the values of r and c in the table? The degrees of freedom can be found using the formula

degrees of freedom $= (r - 1)(c - 1) =$

(b) Go to the row in Table D in the Appendix of the textbook corresponding to the degrees of freedom found in part (a). Where does the value $X^2 = 19.489$ in Table D? What does this tell you about the P-value?

(c) What are the values of r and c in this case? The degrees of freedom can be found using the formula

degrees of freedom $= (r-1)(c-1) =$

Exercise 22.16

KEY CONCEPTS: Chi-square test for goodness of fit

(a) For the professor's grades, fill in the percentages of students earning each grade and compare them to the TA's percentages.

Grade	A	B	C	D/F
Percentage				

(b) Fill in the expected counts for each grade. Using the TA percentages as the null probabilities, the expected count for A's is $np_{10} = 91 \times 0.32 = 29.12$.

Grade	A	B	C	D/F
Expected	29.12			

(c) Complete the calculation of the chi-square goodness of fit statistic. We have indicated how the first term in the sum should be computed:

$$X^2 = \sum \frac{(\text{count of outcome } i - np_{i0})^2}{np_{i0}} = \frac{(22 - 29.12)^2}{29.12} +$$

The degrees of freedom are

$$k - 1 =$$

with k the number of possible outcomes. What can you say about the P-value and what is your conclusion?

Exercise 22.44

KEY CONCEPTS: Two-way tables, testing hypotheses with the X^2 statistic

(a) Think of the rows of the table (disease status) as the "treatments" and the columns (olive oil consumption) as the response; this designation corresponds to how the samples were selected. To be an experiment, what needs to be true about the assignment of the treatments to the subjects? Was this carried out here?

(b) If less than 4% of the cases or controls refused to participate, why would our confidence in the results be strengthened?

(c) The data from the text are reproduced here, with the expected counts printed below the observed counts. Verify a few of the expected counts on your own. If you are not sure how to compute the expected counts, review the complete solution for Exercise 23.7 of this study guide.

Olive Oil

	Low	Medium	High	Total
Colon cancer	398	397	430	1225
	404.39	404.19	416.42	
Rectal cancer	250	241	237	728
	240.32	240.20	247.47	
Controls	1368	1377	1409	4154
	1371.29	1370.61	1412.10	
Total	2016	2015	2076	6107

Is high olive oil consumption more common among patients without cancer than in patients with colon cancer or rectal cancer? Compute the percentages for each of the three below.

Group	Percentage with high olive oil consumption
Colon cancer	
Rectal cancer	
Controls	

What do these percentages suggest?

To do a formal test,

$$X^2 = \sum \frac{(\text{observed} - \text{expected})^2}{\text{expected}} =$$

What is the mean of the X^2 statistic under the null hypothesis? If there is evidence to reject the null hypothesis, the computed value of the statistic should be larger than we would expect it to be under the null hypothesis. Is that true in this case?

What is the P-value? What do you conclude?

COMPLETE SOLUTIONS

Exercise 22.1

(a) The two conditional distributions are given here. Each entry is obtained by dividing the count in the cell by the column total. For example, the percent of Commonwealth students who use Facebook at least once a day is calculated as follows. There are a total of 875 Commonwealth students. Of these, 394 are use Facebook at least once a day, so the percent of Commonwealth students who use Facebook at least once a day is $394/875 = 0.450$.

Campus

Facebook category	University Park	Commonwealth
Do not use Facebook	0.070	0.283
Several times a month or less	0.056	0.087
At least once a week	0.220	0.179
At least once a day	0.654	0.450
Total	1.000	0.999

(b) The bar graph produced by Minitab follows. The key differences are the increased use of Facebook on the University Park campus. This is evidenced by the taller bar for "At least once a day" on the University Park campus and the taller bar fro "Do not use Facebook" on the Commonwealth campus.

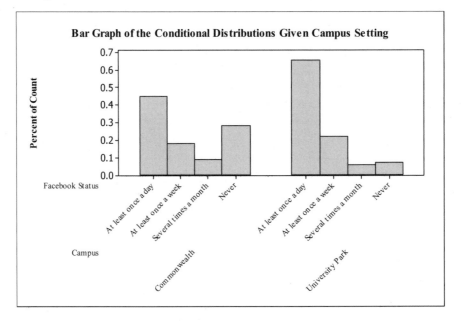

Exercise 22.5

(a) Following is the table of expected counts with the required calculations.

Facebook category	Expected	Observed
Several times a month or less	53.44	76
At least once a week	151.75	157
At least once a day	421.81	394

$$151.75 = \frac{(372)(627)}{1537} \qquad\qquad 421.81 = \frac{(1034)(627)}{1537}$$

The total of the expected counts in this column is 627, which agrees with the total of observed counts except for roundoff error.

(b) There are more in the category "Several times a month" than the null hypothesis calls for. Also, there are slightly more in the"At least once a week category" and fewer in the "At least once a day" category. Review the discussion in part (b) of Exercise 22.1 for a better understanding of this result.

Exercise 22.7

(a) The expected counts are all much larger than 5, so the cell count required for use of chi-square is satisfied.

(b) The null hypothesis is H_0 : there is no relationship between Facebook use and the campus the student attends and the alternative is H_1 : there is some relationship. The value of $X^2 = 19.489$ and the P-value = 0.000. There is very strong evidence of a relationship between Facebook use and the campus the student attends

(c) The largest terms contributing to X^2 are in the Monthly row. This reflects the fact that monthly use is lower among University Park students, and higher among commonwealth students. The third row also has a large contribution to X^2, and this reflects the fact that daily use is higher among University Park students and lower among commonwealth students. See the discussion in part (c) of Exercise 22.1

Exercise 22.11

(a) There are $r = 3$ rows and $c = 2$ columns. Thus, the number of degrees of freedom is

$$(r - 1)(c - 1) = (3 - 1)(2 - 1) = (2)(1) = 2$$

which agrees with the value in the Minitab output.

(b) If we look in the df = 2 row of Table D, we find the following information:

p	.001	.0005
x^*	13.82	15.20

$X^2 = 19.489$ exceeds the entry for $p = .0005$. This tells us that the P-value is less than 0.0005, which agrees with 0.000 reported by Minitab.

(c) In this case there are $r = 4$ rows and $c = 2$ columns. Thus the number of degrees of freedom is

$$(r - 1)(c - 1) = (4 - 1)(2 - 1) = (3)(1) = 3.$$

Exercise 22.16

(a) For the professor's grades, fill in the percentages of students earning each grade and compare them to the TA's percentages.

Grade	A	B	C	D/F
Percentage	22/91 = 0.24	38/91 = 0.42	20/91 = 0.22	11/91 = .0.12

The professor gave a smaller percentage of A's and a higher percentage of D's and F's than the TA. The percentages of B's and C's were similar.

(b) Using the TA percentages as the null probabilities, the expected counts for the professor are

$$np_{10} = 91 \times 0.32 = 29.12$$

$$np_{20} = 91 \times 0.41 = 37.31$$

$$np_{30} = 91 \times 0.20 = 18.20$$

$$np_{40} = 91 \times 0.07 = 6.37$$

Grade	A	B	C	D/F
Expected	29.12	37.31	18.20	6.37

(c) The value of the test statistic is

$$X^2 = \frac{(22-29.12)^2}{29.12} + \frac{(38-37.31)^2}{37.31} + \frac{(20-18.20)^2}{18.20} + \frac{(11-6.37)^2}{6.37} = 1.741 + 0.022 + 0.178 + 3.365 = 5.306$$

The degrees of freedom are $k - 1 = 4 - 1 = 3$. Looking in the df = 3 row of Table D, we find the information

p	.20	.15
x^*	4.64	5.32

giving a P-value between 0.15 and 0.20. There is little evidence that the professor's grade distribution differs from the TA's.

Exercise 22.44

(a) We are thinking of the disease groups as the explanatory variable since the samples were chosen from these groups. The experimenters did not assign the subjects to the disease groups. This makes the data observational, and there is the possibility that differences in olive oil consumption may be a result of confounding variables, not necessarily disease status. (It might be more natural to think of olive oil consumption as the explanatory variable and disease status as the response, but the samples were not selected from the different olive oil consumption groups.)

(b) If there is a high *nonresponse rate*, then it is possible for the percentages in the table to be quite different than the true percentages due to response bias. In this case, the existence of or absence of a relationship could potentially be due to this bias. However, with 96% of the people responding, the bias if it exists could not be large and would have little impact on the results.

(c) The percentages in each group with high olive oil consumptions are computed in the table. There appears to be very little difference in these percentages, suggesting that olive oil in the diet may be unrelated to these forms of cancer.

Group	Percentage with high olive oil consumption
Colon cancer	430/1225 = 0.351
Rectal cancer	237/728 = 0.326
Controls	1409/4154 = 0.339

Using the components of chi-square from Minitab,

$$X^2 = \sum \frac{(\text{observed} - \text{expected})^2}{\text{expected}} = 0.101 + 0.128 + 0.443 + 0.390 + 0.003 + 0.443 +$$

$$0.008 + 0.030 + 0.007 = 1.552$$

If you used the expected counts in the table, which were rounded to two decimals, your answers may disagree slightly with the computations due to rounding error. The mean of the X^2 statistic is equal to the degrees of freedom, which in this case is $(r-1)(c-1) = (3-1)(3-1) = 4$. Since the numerical value of the X^2 statistic is below the mean, there is little evidence to reject the null hypothesis. As can be seen from the table, the observed and expected counts are quite close. From Table D, you can see that the P-value is greater than 0.25. In fact, using statistical software shows that the P-value = 0.817. These data show no evidence of a relationship between disease and olive oil consumption.

CHAPTER 23

INFERENCE FOR REGRESSION

OVERVIEW

In Chapter 5 of the textbook, we first encountered regression. The assumptions that describe the regression model we use in this chapter are the following.

- We have n observations on an explanatory variable x and a response variable y. Our goal is to study or predict the behavior of y for given values of x.

- For any fixed value of x, the response y varies according to a **Normal distribution.** Repeated responses y are **independent** of each other.

- The mean response μ_y has a straight-line relationship with x given by a **population regression line:**

$$\mu_y = \alpha + \beta x$$

The slope β and intercept α are unknown parameters.

- The **standard deviation of y** (call it σ) is the same for all values of x. The value of σ is unknown.

The **true (population) regression line** is $\mu_y = \alpha + \beta x$ and says that the mean response μ_y moves along a straight line as the explanatory variable x changes. The parameters β and α are estimated by the slope b and intercept a of the least-squares regression line, and the formulas for these estimates are

$$b = r \frac{s_y}{s_x}$$

and

$$a = \bar{y} - b\bar{x}$$

where r is the correlation between y and x, $\bar{y}$ is the mean of the y observations, s_y is the standard deviation of the y observations, $\bar{x}$ is the mean of the x observations, and s_x is the standard deviation of the x observations.

The **standard error about the least-squares line** is

$$s = \sqrt{\frac{1}{n-2} \sum \text{residual}^2} = \sqrt{\frac{1}{n-2} \sum (y - \hat{y})^2}$$

where $\hat{y} = a + bx$ is the value we would predict for the response variable based on the least-squares regression line. We use s to estimate the unknown σ in the regression model.

A **level *C* confidence interval** for β is

$$b \pm t^{*}SE_{b}$$

where t^{*} is the critical value for the t distribution with $n-2$ degrees of freedom with area C between $-t^{*}$ and t^{*}, and

$$SE_{b} = \frac{s}{\sqrt{\Sigma(x-\bar{x})^{2}}}$$

is the standard error of the least-squares slope b. SE_{b} is usually computed using a calculator or statistical software.

The **test of the hypothesis** H_{0}: $\beta = 0$ is based on the t statistic

$$t = \frac{b}{SE_{b}}$$

with P-values computed from the t distribution with $n-2$ degrees of freedom. This test is also a test of the hypothesis that the correlation is 0 in the population.

A **level *C* confidence interval for the mean response** μ_{y} when x takes the value x^{*} is

$$\hat{y} \pm t^{*}SE_{\hat{\mu}}$$

where $\hat{y} = a + bx$, t^{*} is the critical value for the t distribution with $n-2$ degrees of freedom and area C between $-t^{*}$ and t^{*} and

$$SE_{\hat{\mu}} = s\sqrt{\frac{1}{n} + \frac{(x^{*}-\bar{x})^{2}}{\Sigma(x-\bar{x})^{2}}}$$

$SE_{\hat{\mu}}$ is usually computed using a calculator or statistical software.

A **level *C* prediction interval for a single observation** on y when x takes the value x^{*} is

$$\hat{y} \pm t^{*}SE_{\hat{y}}$$

where t^{*} is the critical value for the t distribution with $n-2$ degrees of freedom and and area C between $-t^{*}$ and t^{*} and

$$SE_{\hat{y}} = s\sqrt{1 + \frac{1}{n} + \frac{(x^{*}-\bar{x})^{2}}{\Sigma(x-\bar{x})^{2}}}$$

$SE_{\hat{y}}$ is usually computed using a calculator or statistical software.

Finally, it is always good practice to check that the data satisfy the linear regression model assumptions before doing inference. Scatterplots and residual plots are useful tools for checking these assumptions.

GUIDED SOLUTIONS

Exercise 23.1

KEY CONCEPTS: Scatterplots, correlation, linear regression, residuals, standard error of the least-squares line

(a) First, examine the data and judge whether the relationship between "Distance" and "Days" is positive or negative. Sketch your scatterplot on the axes provided, or use software.

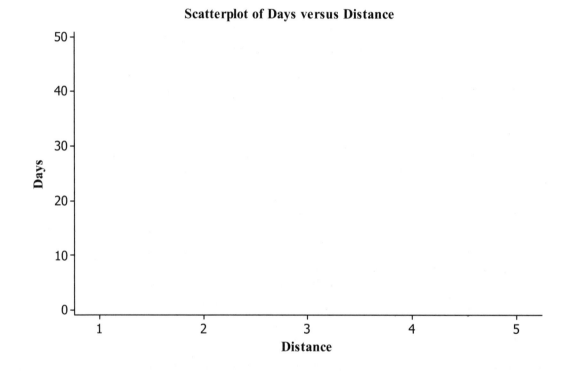

Use your calculator (or statistical software) to compute the correlation r:

$r =$

(b) What does the slope β of the true regression line say about the number of days until group infection and a group's distance from the first infected group?

Enter your estimates of the slope β and intercept α of the true regression line. Use software or your calculator, or compute these values manually using the formulas in Chapter 6 of your textbook.

Estimate of $\beta =$

Estimate of $\alpha =$

Although it isn't asked for in this part, write the equation of the least-squares regression line for predicting the number of days to infection for a gorilla group given its distance from the first group infected. You'll use this in part (c)

The least-squares regression line is: $\hat{y} =$

(c) To compute the residuals, complete the table. Remember, to compute the predicted number of days until infection, use the least-squares regression line.

Distance from first group infected	Predicted number of days until infection	Residual (prediction error)
1		
3		
4		
4		
4		
5		

Compute the sum of residuals (sum of prediction errors). They should sum to zero.

$$\sum \text{residual} =$$

Now estimate the standard deviation σ by computing

$$\sum \text{residual}^2 =$$

and then completing the following calculation. This is an estimate of σ.

$$s = \sqrt{\frac{1}{n-2}\sum \text{residual}^2} =$$

Exercise 23.4

KEY CONCEPTS: Tests for the slope of the least-squares regression line

(a)The test of the hypotheses $H_0 : \beta = 0$ is based on the t statistic $t = \dfrac{b}{SE_b}$.

In the statement of the problem, we are told that $b = 11.263$ and $SE_b = 1.591$. The value of b is slightly different than the value we found in Exercise 23.1, due to differences in how much rounding was done at intermediate stages of the calculations.

Compute the test statistic:

$$t = \frac{b}{SE_b} =$$

(b) What are the degrees of freedom for t? Refer to the original data in Exercise 23.1 of your textbook to determine the sample size n.

Degrees of freedom $= n - 2 =$

Now, use Table C to estimate the P-value for testing with the alternative hypothesis $H_a : \beta > 0$, which hypothesizes a positive linear association between "Days" and "Distance."

P-value:

What do you conclude?

Exercise 23.38

KEY CONCEPTS: Scatterplots, examining residuals, confidence intervals for the slope

(a) Use software or a calculator to compute the correlation between "Time" and "Calories":

Use software or a calculator to compute the equation of the least-squares regression line. Don't forget to have the computer or your calculator save the residuals, as we'll use them in part (b):

$\hat{y} =$

Use software or the axes provided to make a scatterplot of Calories versus Time.

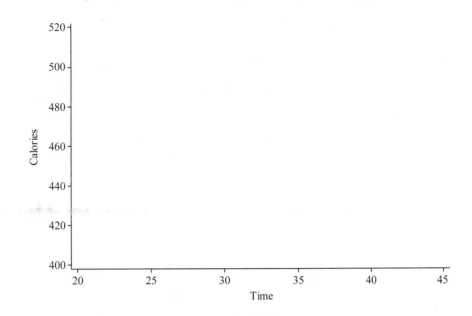

(b) Here, we'll check conditions needed for regression inference.

First, to check for a Linear Relationship, and to check whether spread about the line stays the same for all values of the explanatory variable, plot the residuals against Time (the explanatory variable):

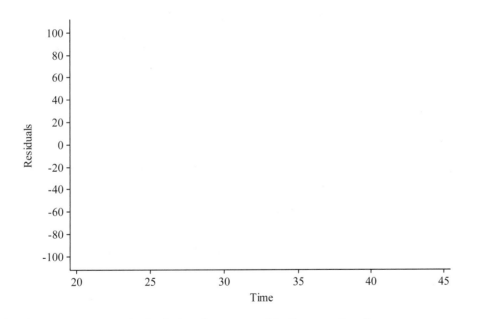

Does this plot show any systematic deviation from a roughly linear pattern?

Does this plot show any systematic change in spread as "Time" changes?

Are the observations independent? Is this obvious?

Finally, look for evidence that the variation about the line appear to be Normal. Use software or the axes that follow (with class intervals $-40 \leq$ residual < -30, $-30 \leq$ residual < -20, $-20 \leq$ residual < -10, and so on) to make a histogram.

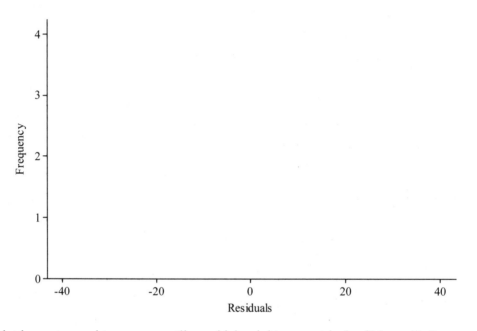

Does this plot have strong skewness or outliers which might suggest lack of Normality?

(c) In this problem, the rate of change in calories consumed as time at the table increases is the slope of the population line, β. Hence, we need to construct a 95% confidence interval for β. Recall that a level C confidence interval for β is

$$b \pm t^* \text{SE}_b$$

where t^* is the critical value for the t distribution with $n - 2$ degrees of freedom with area C between $-t^*$ and t^*, and

$$\text{SE}_b = \frac{s}{\sqrt{\sum (x - \bar{x})^2}}$$

is the standard error of the least-squares slope b.

In this exercise, b and SE_b can be read directly from the output of statistical software. Record their values.

$b =$

$SE_b =$

Now, find t^* for a 95% confidence interval from Table C (what is n here?).

$t^* =$

Compute the 90% confidence interval:

Interpret this confidence interval in the context of this problem.

Exercise 23.40

KEY CONCEPTS: Prediction, prediction intervals

We used Minitab to compute a prediction of Calories when Time = 40. The output follows:

```
The regression equation is
Calories = 561 - 3.08 Time

Predictor        Coef        Stdev      t-ratio          p
Constant        560.65       29.37        19.09      0.000
Time           -3.0771       0.8498       -3.62      0.002

s = 23.40       R-sq = 42.1%      R-sq(adj) = 38.9%

Analysis of Variance

SOURCE         DF          SS          MS          F          p
Regression      1       7177.6      7177.6      13.11      0.002
Error          18       9854.4       547.5
Total          19      17032.0

     Fit   Stdev.Fit        95.0% C.I.              95.0% P.I.
  437.57        7.30    ( 422.23,   452.91)    ( 386.06,   489.08)
```

Where in this output does one find the 95% confidence interval to predict Rachel's calorie consumption at lunch? Refer to Examples 23.7 and 23.8 in the textbook if you need help.

95% prediction interval:

COMPLETE SOLUTIONS

Exercise 23.1

(a) If we look at the data, we see that as a gorilla group's distance from the first infection increases, so does the number of days until that group is infected. Thus, there is a positive association between "Days" and "Distance." A scatterplot of the data with price as the explanatory variable follows.

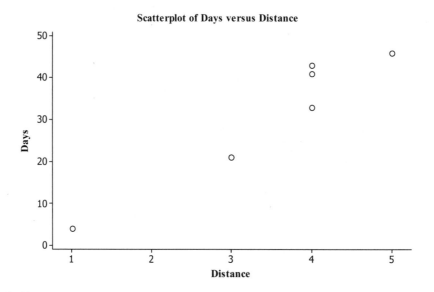

The scatterplot indicates a strong positive linear association between "Distance" and "Days."
The correlation r is given by $r = 0.962$. This is consistent with the scatterplot as suggesting a strong linear relationship between "Distance" and "Days."

The estimate of β is $b = 11.3$ days per distance unit.
The estimate of α is $a = -8.09$ days.

The equation of the least-squares regression line for predicting days to infection for a gorilla group given its distance from the initial group infected is: $\text{Days} = -8.09 + 11.3 \times \text{Distance}$

(b) The slope of the population regression line, β, is the number of additional days (on average) required to infect a gorilla group one additional distance unit from the original infection group. You might think of this as a measure of the rate of the infection's spread - on average it takes β days for the infection to spread to an additional home range. The estimate of β is $b = 11.3$ days per distance unit.

The estimate of α is $a = -8.09$ days.

The equation of the least-squares regression line for predicting days to infection for a gorilla group given its distance from the initial group infected is: $\text{Days} = -8.09 + 11.3 \times \text{Distance}$

(c) The residuals for the six data points are given in the table.

Distance from first group infected	Predicted number of days until infection	Residual (prediction error)
1	3.18	$4 - 3.18 = 0.82$
3	25.70	$21 - 25.70 = -4.70$
4	36.96	$33 - 36.96 = -3.96$
4	36.96	$41 - 36.96 = 4.04$
4	36.96	$43 - 36.96 = 6.04$
5	48.23	$46 - 48.23 = -2.23$

The sum of the residuals listed is $\sum \text{residual} = 0.01$. The difference from 0 is due to rounding in the parameter estimates above. To estimate the standard deviation σ in the regression model, we first calculate the sum of the squares of the residuals listed:

$$\sum \text{residual}^2 = 0.82^2 + (-4.70)^2 + \cdots + (-2.23)^2 = 96.22.$$

Our estimate of the standard deviation σ in the regression model is therefore

$$s = \sqrt{\frac{1}{n-2} \sum \text{residual}^2} = \sqrt{\frac{1}{6-2}(96.22)} = 4.90 \text{ days}.$$

Exercise 23.4

(a) $b = 11.263$ and $SE_b = 1.591$, so $t = \dfrac{b}{SE_b} = \dfrac{11.263}{1.591} = 7.079$

(b) Referring to the original data in Exercise 23.1 of the textbook, we see that $n = 6$.

Degrees of freedom $= n - 2 = 6 - 2 = 4$

To estimate the P-value, we use Table C with df $= 4$ and refer to the P-values corresponding to the two values of t^* that bracket the computed value of $t = 7.079$:

t^*	5.598	7.173
One-sided P	.0025	.001

Because the test is two-sided, $0.001 < P\text{-value} < 0.0025$. Statistical software (Minitab) gives a P-value of 0.002. There is extremely strong (overwhelming) evidence to support a positive linear association between distance of a gorilla group from the primary infection group and the number of days it takes for the infection to reach the group.

Exercise 23.38

(a) Here is a scatterplot showing the relationship between time at the table and calories consumed.

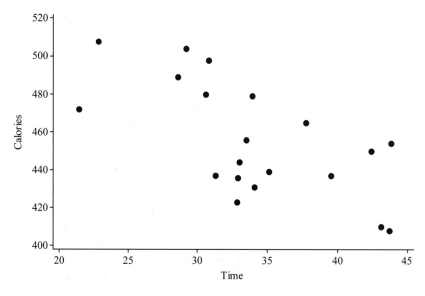

The correlation between "Calories" and "Time" is $r = -0.649$.

The overall pattern is roughly (perhaps weakly) linear with a negative slope. There are no clear outliers or strongly influential data points, it seems.

Using statistical software, we find that the equation of the least-squares line is

$$\hat{y} = 560.65 - 3.08 \times \text{time}$$

(b) A scatterplot of the residuals against "Time" follows.

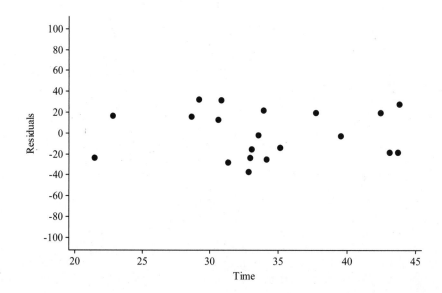

This plot is useful for addressing the first two of the four conditions we check:

Does the relationship appear linear?

This scatterplot magnifies deviations from the regression line, making it easier to detect any non-linear pattern in the data. Based on this plot, there is little reason to doubt that the relationship between "Calories" and "Time" is linear.

Does the spread about the line stay the same?

The scatterplot of residuals versus "Time" seems to suggest that the spread about the line is roughly constant. Points seem to lie consistently in a band between –40 and +40.

Are the observations independent?

The answer is not clear. These are observations on 20 different children rather than on a single child, and that is good. However, we do not know if the children were selected at random. In addition, we do not know if the children were all together so that the behavior of one child could influence the behavior of another. Are there children from the same family in this group? These issues would impact independence of observations.

Does the variation about the line appear to be Normal?

The histogram that follows has a gap and is not particularly bell-shaped. On the other hand there do not appear to be any outliers or extreme skew. With only 20 observations, it's difficult to assess non-Normality here.

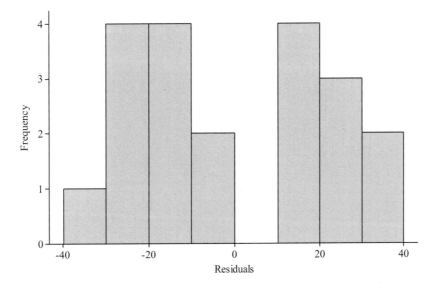

The conditions for inference (for a sample of size 20) are approximately satisfied.

(c) From statistical software, we find that

$$b = -3.08$$

$$SE_b = 0.85$$

For a 95% confidence interval from Table C with $n = 20$ (and $n - 2 = 18$),

$$t^* = 2.101$$

We use these to compute the 95% confidence interval for the true slope of the regression line:

$$b \pm t^*SE_b = -3.08 \pm (2.101)(0.85) = -3.08 \pm 1.79$$
$$\text{or } -4.87 \text{ to} -1.29 \text{ calories per minute.}$$

With 95% confidence, each minute spent at the table reduces calories consumed by between 1.29 calories and 4.87 calories.

Exercise 23.40

Using software (Minitab, in this case):

The output from Minitab follows:

```
The regression equation is
Calories = 561 - 3.08 Time

Predictor        Coef        Stdev      t-ratio         p
Constant       560.65        29.37        19.09     0.000
Time          -3.0771        0.8498       -3.62     0.002

s = 23.40        R-sq = 42.1%      R-sq(adj) = 38.9%

Analysis of Variance

SOURCE         DF          SS           MS          F         p
Regression     1        7177.6       7177.6      13.11     0.002
Error          18       9854.4        547.5
Total          19      17032.0

     Fit  Stdev.Fit       95.0% C.I.           95.0% P.I.
   437.57       7.30   (422.23,  452.91)   (386.06,  489.08)
```

The "Fit" entry gives the predicted calories. Minitab gives both the 95% confidence interval for the mean response and the prediction interval for a single observation. We are predicting a single observation, so the column labeled "95% PI" contains the interval we want. We see that this 95% prediction interval is (386.06, 489.08). With 95% confidence, the mean number of calories consumed by Rachel at lunch is between 386 and 489 calories, roughly.

CHAPTER 24

ONE-WAY ANALYSIS OF VARIANCE: COMPARING SEVERAL MEANS

OVERVIEW

The two-sample t procedures compare the means of two populations. However, when the mean is the best description of the center of a distribution, we may want to compare several population means or several treatment means in a designed experiment. For example, we might be interested in comparing the mean weight loss by dieters on three different diet programs or the mean yield of four varieties of green beans.

The method we use to compare more than two population means is the **analysis of variance (ANOVA) F test.** This test is also called the **one-way ANOVA.** The ANOVA F test is an overall test that looks for any difference between a group of I means. The null hypothesis is $H_0 : \mu_1 = \mu_2 = \cdots = \mu_I$, where we tell the population means apart by using the subscripts 1 through I. The alternative hypothesis is H_a: not all the means are equal. In a more advanced course, you would study formal inference procedures for a follow-up analysis to decide which means differ and to estimate how large the differences are. Note that formally the ANOVA F test is a different test from the F test introduced in Chapter 19 of your text that compared the standard deviations of two populations, although the ANOVA F test does involve the comparison of two measures of variation.

The ANOVA F test compares the variation among the groups to the variation within the groups through the **F statistic,**

$$F = \frac{\text{variation among the sample means}}{\text{variation among the individuals in the same sample}}$$

The important thing to take away from this chapter is the rationale behind the ANOVA F test. The particulars of the calculation are not as important since software usually calculates the numbers for us.

The F statistic has the F distribution. The distribution is completely defined by its two degrees of freedom parameters, the numerator degrees of freedom and the denominator degrees of freedom. The numerator has $I - 1$ degrees of freedom, where I is the number of populations we are comparing. The denominator has $N - I$ degrees of freedom, where N is the total number of observations. The F distribution is usually written $F(I - 1, N - I)$.

We make the following assumptions for ANOVA:

- There are I independent SRSs.
- Each population is Normally distributed with its own mean, μ_i.
- All populations have the same standard deviation, σ.

The first assumption is the most important. The test is robust against non-Normality, but it is still important to check for outliers and/or skewness that would make the mean a poor measure of the center of the distribution. As for the assumption of equal standard deviations, make sure that the largest sample standard deviation is no more than twice the smallest standard deviation.

Although it is generally best to leave the ANOVA computations to statistical software, seeing the formulas sometimes helps one to obtain a better understanding of the procedure. In addition, there are times when the original data are not available and you have only the group means and standard deviations or standard error. In these instances, the formulas described here are required to carry out the ANOVA F test.

The F statistic is $F = \dfrac{\text{MSG}}{\text{MSE}}$, where MSG is the **mean square for groups,**

$$\text{MSG} = \frac{n_1(\overline{x}_1 - \overline{x})^2 + n_2(\overline{x}_2 - \overline{x})^2 + \cdots + n_I(\overline{x}_I - \overline{x})^2}{I - 1}$$

with

$$\overline{x} = \frac{n_1\overline{x}_1 + n_2\overline{x}_2 + \cdots + n_I\overline{x}_I}{N}$$

and MSE is the **error mean square,**

$$\text{MSE} = \frac{s_1^2(n_1 - 1) + s_2^2(n_2 - 1) + \cdots + s_I^2(n_I - 1)}{N - I}.$$

Because MSE is an average of the individual sample variances, it is also called the **pooled sample variance,** written s_P^2, and its square root, $s_p = \sqrt{\text{MSE}}$ is called the **pooled standard deviation.** We can also make a confidence interval for any of the means by using the formula $\overline{x}_i \pm t^* \dfrac{s_p}{\sqrt{n_i}}$. The critical value is t^* from the t distribution with $N - I$ degrees of freedom.

GUIDED SOLUTIONS

Exercise 24.3

KEY CONCEPTS: Side-by-side stemplots, ANOVA hypotheses, drawing conclusions from ANOVA output

(a) Complete the stemplots on the next page (they use split stems). From the stemplots, would you say that any of the groups show outliers or extreme skewness? What effects of logging are visible from the stemplots?

Never logged	Logged 1 year ago	Logged 8 years ago
0	0	0
0	0	0
1	1	1
1	1	1
2	2	2
2	2	2
3	3	3

(b) What do the means suggest about the effect of logging?

(c) State the null and alternative hypotheses, letting μ_1, μ_2, and μ_3 denote the means for the three groups.

H_0: H_a:

From the output, determine the values of the ANOVA F statistic and its P-value.   What are your conclusions?

F statistic = P-value =

Exercise 24.10

KEY CONCEPTS: ANOVA degrees of freedom, computing P-values from Table D

(a) In the table, fill in the numerical values and explain in words the meaning of each symbol we are using in the notation for the one-way ANOVA. Group 1, group 2, and group 3 are identified in the exercise.

Symbol value	Verbal meaning
I	
n_1	
n_2	
n_3	
N	

(b) Use the text formulas and the results from part (a) to give the numerator and denominator degrees of freedom. Check your answers against the Excel output given in Exercise 24.4.

Numerator degrees of freedom =

Denominator degrees of freedom =

The value $F = 11.43$ needs to be referred to an $F(2, 30)$ distribution. What can you say about the P-value from Table D?

Exercise 24.13

KEY CONCEPTS: Checking standard deviations, ANOVA computations

(a) Do the standard deviations satisfy the rule of thumb for using ANOVA?

$$\frac{\text{largest sample standard deviation}}{\text{smallest sample standard deviation}} =$$

(b) You will need the means, sample sizes and standard deviations for the three groups to do the calculations. To compute MSG, you first need to compute the overall mean

$$\bar{x} = \frac{n_1 \bar{x}_1 + n_2 \bar{x}_2 + \cdots + n_I \bar{x}_I}{N} =$$

and then substitute the means, sample sizes, and overall mean into the formula

$$\text{MSG} = \frac{n_1 (\bar{x}_1 - \bar{x})^2 + n_2 (\bar{x}_2 - \bar{x})^2 + \cdots + n_I (\bar{x}_I - \bar{x})^2}{I - 1} =$$

MSE is then obtained from the formula

$$\text{MSE} = \frac{s_1^2 (n_1 - 1) + s_2^2 (n_2 - 1) + \cdots + s_I^2 (n_I - 1)}{N - I} =$$

The F statistic is calculated as

$$F = \frac{\text{MSG}}{\text{MSE}} =$$

(c) What are the degrees of freedom for the ANOVA F statistic? This is the distribution you use to find the P-value in Table D.

The P-value is 0.634. Is there evidence that mean weight losses of people who follow the three exercise programs differ?

Exercise 24.34

KEY CONCEPTS: Four-step rule, one-way analysis of variance

The four-step process follows:

State. What is the practical question that requires a statistical test?

Plan. Identify the parameters, state null and alternative hypotheses, and choose appropriate test.

Solve. (1) Check the conditions, (2) calculate the test statistic, and (3) find the *P*-value.

Conclude. Return to the practical question to describe your results in this setting.

To apply the steps to this problem, here are some suggestions. You may want to use Example 24.4 of your text as a guide.

State. Using the language in the problem, state the goals of the study. Since the data set is small, you can include the data as well.

Plan. Be sure to check to make sure that you can safely use ANOVA.

Solve. The output here is from MINITAB. If you are using software for your course, you should try to run the one-way ANOVA for this problem on your own software. Carrying out the calculations by hand is quite tedious even for a small data set such as this one.

One-way ANOVA: Breaking Strength versus Weeks

```
Source  DF      SS      MS    F      P
Weeks    2   381.7   190.9  3.70  0.056
Error   12   619.2    51.6
Total   14  1000.9

S = 7.183   R-Sq = 38.14%   R-Sq(adj) = 27.83%
```

```
                          Individual 95% CIs For Mean Based on
                          Pooled StDev
Level   N     Mean   StDev   ---+---------+---------+---------+------
2       5   123.80    4.60   (---------*---------)
4       5   123.60    6.54   (---------*---------)
8       5   134.40    9.53                 (---------*---------)
                             ---+---------+---------+---------+------
                          119.0       126.0     133.0     140.0

Pooled StDev = 7.18
```

Conclude. Be careful with your conclusion. Even though the data shows some evidence of a difference in breaking strength over weeks, do the data support the conjecture that polyester loses strength over time?

COMPLETE SOLUTIONS

Exercise 24.3

(a) The side-by-side stemplots are completed here. Extreme skewness is not evident. There is a low outlier in the "Logged 8 years ago" column. The counts of trees in the plots that were never logged appear to be larger than those that were logged in the stemplots. There appears to be little difference between the counts from plots logged 1 year ago and plots logged 8 years ago.

Never logged	Logged 1 year ago	Logged 8 years ago
0	0 | 2	0 | 4
0	0 | 9	0
1	1 | 2 2 4 4	1 | 2 2
1 | 6 9 9	1 | 5 7 7 8 9	1 | 5 8 8 9
2 | 0 1 2 4	2 | 0	2 | 2 2
2 | 7 7 8 9	2	2
3 | 3	3	3

(b) The means suggest that logging reduces the number of trees per plot and that the recovery may be quite slow as there is little difference in the means for the logged 1 year ago and logged 8 years ago groups.

(c) The hypotheses are $H_0 : \mu_1 = \mu_2 = \mu_3$ and H_a: not all of μ_1, μ_2, and μ_3 are equal.

The overall ANOVA F test has $F = 11.4257$ and P-value $= 0.000205$, so there is strong evidence of a difference in mean trees per plot among the three groups. The ANOVA F test does not tell us which groups are different, but examination of the means and stemplots shows both the logged groups to have lower mean trees per plot than the never logged group, though there appears to be little difference between the plots logged 1 year ago and the plots logged 8 years ago.

Exercise 24.10

(a) The table gives the value and states in words the meaning of each symbol used in a one-way ANOVA.

Symbol	Value	Verbal meaning
I	3	Number of groups
n_1	12	Number of plots in the never logged group
n_2	12	Number of plots in the logged 1 year ago group
n_3	9	Number of plots in the logged 8 years ago group
N	33	Total number of plots in the experiment

(b) The ANOVA F statistic has the F distribution with $I - 1 = 3 - 1 = 2$ degrees of freedom in the numerator and $N - I = 33 - 3 = 30$ degrees of freedom in the denominator.

The critical value of 9.22 corresponds to a tail probability of 0.001 for an $F(2, 25)$ distribution. Since the critical value for an $F(2, 30)$ distribution corresponding to a tail area of 0.001 would be even smaller, the value $F = 11.4257$ would exceed this value. We can say that the P-value is less than 0.001, which agrees with the computer output.

Exercise 24.13

(a) The ratio of the largest to the smallest standard deviations is

$$\frac{\text{largest sample standard deviation}}{\text{smallest sample standard deviation}} = \frac{5.2}{4.2} = 1.24 < 2$$

so the rule of thumb for safe use of ANOVA is satisfied.

(b) We have three treatments, so $I = 3$ and the formula for the overall mean gives

$$\bar{x} = \frac{n_1\bar{x}_1 + n_2\bar{x}_2 + \cdots + n_I\bar{x}_I}{N} = \frac{n_1\bar{x}_1 + n_2\bar{x}_2 + \cdots + n_I\bar{x}_I}{n_1 + n_2 + n_3} = \frac{(37)(10.2)+(36)(9.3)+(42)(10.2)}{37+36+42} = \frac{1140.6}{115} = 9.9183$$

and then substituting the means, sample sizes, and overall mean into the formula

$$\text{MSG} = \frac{n_1(\bar{x}_1 - \bar{x})^2 + n_2(\bar{x}_2 - \bar{x})^2 + \cdots + n_I(\bar{x}_I - \bar{x})^2}{I - 1}$$

$$= \frac{37(10.2 - 9.9183)^2 + 36(9.3 - 9.9183)^2 + 42(10.2 - 9.9183)^2}{3 - 1} = \frac{20.0317}{2} = 10.016$$

Remembering to square the standard deviations, MSE is then obtained from the formula

$$\text{MSE} = \frac{s_1^2(n_1 - 1) + s_2^2(n_2 - 1) + \cdots + s_I^2(n_I - 1)}{N - I}$$

$$= \frac{(4.2)^2(37 - 1)+(4.5)^2(36 - 1)+(5.2)^2(42 - 1)}{115 - 3} = \frac{2452.43}{112} = 21.897.$$

The *F* statistic is calculated as

$$F = \frac{MSG}{MSE} = \frac{10.016}{21.897} = 0.457.$$

(c) The numerator has $I - 1 = 3 - 1 = 2$ degrees of freedom and the denominator has $N - I = 115 - 3 = 112$ degrees of freedom, so the F(2, 112) distribution is used to find the *P*-value. The *P*-value of 0.634 from software tells us that there is no evidence of a difference in weight loss after six months among the three treatments.

Exercise 24.34

State. To study the rate of decay of polyester in landfills, a researcher buried strips of polyester in soil for different lengths of time, then dug up the strips and measured the force required to break them. Breaking strength was chosen as it is easy to measure and should be a good indicator of decay with lower breaking strength indicating greater decay. Fifteen strips were buried in well drained soil and 5 strips, chosen at random, were dug up after 2, 4, and 8 weeks. The breaking strengths in pounds follow.

Breaking Strength-2 Weeks	Breaking Strength-4 Weeks	Breaking Strength-8 Weeks
118	130	122
126	120	136
126	114	128
120	126	146
129	128	140

Plan. The ratio of the largest to the smallest standard deviations is

$$\frac{\text{largest sample standard deviation}}{\text{smallest sample standard deviation}} = \frac{9.53}{4.60} = 2.07$$

which is slightly larger than 2. The rule of thumb is conservative, and with equal sample sizes in the three groups many statisticians would proceed with the ANOVA in this situation. The one-way ANOVA will be used to determine if there is evidence of a difference in mean breaking strength among polyester strips buried in soil for 2, 4, or 8 weeks.

Solve. The following output is from MINITAB. The first thing to notice is that the sample mean breaking strengths are 123.80 for the two-week treatment, 123.60 for the four-week treatment and 134.40 for the eight-week treatment. The *P*-value is 0.056, which provides slight evidence of a difference in breaking strengths for the three groups. However, the ANOVA does not demonstrate that polyester is losing strength over the time period studied.

One-way ANOVA: Breaking Strength versus Weeks

```
Source   DF      SS     MS     F      P
Weeks     2   381.7  190.9  3.70  0.056
Error    12   619.2   51.6
Total    14  1000.9

S = 7.183   R-Sq = 38.14%   R-Sq(adj) = 27.83%
```

```
                             Individual 95% CIs For Mean Based on
                             Pooled StDev
Level   N     Mean    StDev   ---+---------+---------+---------+------
2       5   123.80     4.60   (---------*---------)
4       5   123.60     6.54   (---------*---------)
8       5   134.40     9.53                   (---------*---------)
                             ---+---------+---------+---------+------
                             119.0     126.0     133.0     140.0
```

Pooled StDev = 7.18

Conclude. Since it seems unlikely that polyester could be getting stronger over time, we would consider that the difference in sample means, although somewhat large, can be explained by chance despite the *P*-value of 0.056. The explanation would be that decay did not occur over 8 weeks and some of the stronger strips ended up in the 8-week group just by chance. Since the study may not have been carried out over a long enough period of time to see an effect of time on breaking strength, further experiments over longer time periods may need to be run.

CHAPTER 25

NONPARAMETRIC TESTS

OVERVIEW

Many of the statistical procedures described in previous chapters assume that samples are drawn from Normal populations. **Nonparametric tests** do not require any specific form for the distributions of the populations from which the samples are drawn. Many nonparametric tests are **rank tests;** that is, they are based on the **ranks** of the observations rather than on the observations themselves. When ranking the observations from smallest to largest, tied observations receive the average of their ranks.

The **Wilcoxon rank sum test** compares two distributions. The objective is to determine if one distribution has systematically larger values than the other. The observations are ranked, and the **Wilcoxon rank sum statistic** W is the sum of the ranks of one of the samples. The Wilcoxon rank sum test can be used in place of the **two-sample t test** when samples are small or the populations are far from Normal.

Exact P-values for the Wilcoxon rank sum test require special tables and are produced by some statistical software. However, many statistical software packages give only approximate P-values based on a Normal approximation, typically with a continuity correction. Many packages also make an adjustment in the Normal approximation when there are ties in the ranks.

The **Wilcoxon signed rank test** is a nonparametric test for matched pairs. It tests the null hypothesis that there is no systematic difference between the observations within a pair against the alternative that one observation tends to be larger.

The test is based on the **Wilcoxon signed rank statistic W^+**, which provides another example of a nonparametric test using ranks. The absolute values of the differences between matched pairs of observations are ranked and the sum of the ranks of the positive (or negative) differences gives the value of W^+. The **matched pairs t test** is an alternative test that assumes a Normal distribution for the differences.

P-values for the signed rank test can be found in special tables of the distribution or a Normal approximation to the distribution of W^+. Some software computes the exact P-value and other software uses the Normal approximation, typically with a continuity correction. Many packages make an adjustment in the Normal approximation when there are ties in the ranks.

The **Kruskal-Wallis test** is the nonparametric test for the **one-way analysis of variance** setting. In comparing several populations, it tests the null hypothesis that the distribution of the response variable is the same in all groups and the alternative hypothesis that some groups have distributions of the response variable that are systematically larger than others.

The **Kruskal-Wallis statistic** H compares the average ranks received for the different samples. If the alternative is true, some should be larger than others. Computationally, it essentially arises from performing the usual one-way ANOVA to the ranks of the observations rather than the observations themselves.

P-values for the Kruskal-Wallis test can be found in special tables of the distribution or a chi-square approximation to the distribution of H. When the sample sizes are not too small, the distribution of H for comparing I populations has approximately a chi-square distribution with $I-1$ degrees of freedom. Some software computes the exact P-value and other software uses the chi-square approximation, typically with an adjustment in the chi-square approximation when there are ties in the ranks.

GUIDED SOLUTIONS

Exercise 25.12

KEY CONCEPTS: Ranking data, two-sample problem, Wilcoxon rank sum test

(a) Order the observations from smallest to largest. Use a different color for or underline observations in the supplemented group to make it easier to determine the ranks assigned to each group.

(b) Suppose the first sample is the supplemented group and the second sample is the control group. The choice of which sample we call the first sample and which we call the second sample is arbitrary. However, the Wilcoxon rank sum test is the sum of the ranks of the first sample, and the formulas for the mean and variance of W distinguish between the sample sizes for the first and the second samples. Use the ranks of the supplemented group to compute the value of W.

$$W =$$

(c) What are the values of n_1, n_2, and N? Use these values to evaluate the mean and standard deviation of W according to the formulas that follow:

$$\mu_W = \frac{n_1(N+1)}{2} =$$

$$\sigma_W = \sqrt{\frac{n_1 n_2 (N+1)}{12}} =$$

Use the mean and standard deviation to compute the standardized rank sum statistic:

$$z = \frac{W - \mu_W}{\sigma_W} =$$

What kind of values would W have if the alternative were true? Use the Normal approximation to find the approximate P-value. If you have access to software or tables to evaluate the exact P-value, compare it with the approximation.

P-value =

What are your conclusions?

Exercise 25.25

KEY CONCEPTS: Matched pairs, Wilcoxon signed rank statistic

(a) First, give the null and alternative hypotheses. If the cola loses sweetness, what will be the sign of the sweetness loss (sweetness before storage minus sweetness after storage)?

H_0:
H_a:

To compute the Wilcoxon signed rank statistic, order the absolute values of the differences and rank them. When there are ties, be careful computing the ranks. In any tied group of observations, each observation should each receive the average rank for the group. (Note that the negative observations are in bold and italics.) The ranks of the two smallest absolute values are given to help get you started. Now, fill in the remaining ranks.

Absolute values	Ranks
0.4	1.5
0.4	1.5
0.7	
1.1	
1.2	
1.3	
2.0	
2.0	
2.2	
2.3	

To see how the ranks are computed, the 0.4's would get ranks 1 and 2, so their average rank is 1.5. The 0.7 would get rank 3 and so on. If W^+ is the sum of the ranks of the positive observations, compute the value of W^+.

$$W^+ =$$

Evaluate the mean and standard deviation of W^+ according to the following formulas:

$$\mu_{W^+} = \frac{n(n+1)}{4} =$$

$$\sigma_{W^+} = \sqrt{\frac{n(n+1)(2n+1)}{24}} =$$

Now, use the mean and standard deviation to compute the standardized rank sum statistic:

$$z = \frac{W^+ - \mu_{W^+}}{\sigma_{W^+}} =$$

Do you expect W^+ to be small or large if the alternative is true? Use the Normal approximation to find the approximate P-value.

What are your conclusions?

(b) How do the P-values from the Wilcoxon signed rank test and the one-sample t test compare?

For the one-sample t test, give the null and alternative hypotheses.

H_0:
H_a:

What are the assumptions for each of the procedures?

Exercise 25.48

KEY CONCEPTS: One-way ANOVA, Kruskal-Wallis statistic

We are going to use the Kruskal-Wallis test to determine if nematodes in soil affect plant growth. First, give the null and alternative hypotheses for the Kruskal-Wallis test.

H_0:

H_a:

To compute the Kruskal-Wallis test statistic, the 16 observations are first arranged in increasing order as follows, where we have kept track of the group for each observation. Fill in the ranks. Remember that there is one tied observation.

```
Growth   3.2     4.6     5.0     5.3     5.4     5.8     7.4
Group  10000    5000    5000   10000    5000   10000    5000
Rank

Growth   7.5     8.2     9.1     9.2    10.8    11.1    11.1
Group  10000    1000       0       0       0    1000    1000
Rank

Growth  11.3    13.5
Group   1000       0
Rank
```

Fill in the following table, which gives the ranks for each of the nematode groups and the sum of ranks for each group.

Nematodes	Ranks	Sum of ranks
0		
1000		
5000		
10000		

Use the sum of ranks for the four groups to evaluate the Kruskal-Wallis statistic. What are the numerical values of n_i and N in the formula?

$$H = \frac{12}{N(N+1)}\sum \frac{R_i^2}{n_i} - 3(N+1) =$$

The value of H is compared with critical values in Table E for a chi-square distribution with $I - 1$ degrees of freedom, where I is the number of groups. What is the P-value and what do you conclude?

COMPLETE SOLUTIONS

Exercise 25.12

(a) First, the observations are ordered from smallest to largest. The observations given in bold are from the supplemented group.

Observations	Ranks
−1.2	1
2.3	2
4.6	3.5
4.6	3.5
5.4	5
6.0	6
7.7	7.5
7.7	7.5
11.3	9.5
11.3	9.5
11.4	11
15.5	12
16.5	13

(b) The Wilcoxon rank sum statistic is

$$W = 5 + 7.5 + 9.5 + 9.5 + 11 + 12 + 13 = 67.5$$

(c) The sample sizes are $n_1 = 7$, $n_2 = 6$, and $N = 13$. The values for the mean and variance are

$$\mu_W = \frac{n_1(N+1)}{2} = \frac{7(13)}{2} = 45.5$$

and

$$\sigma_W = \sqrt{\frac{n_1 n_2 (N+1)}{12}} = \sqrt{\frac{(7)(6)(13)}{12}} = 6.745$$

and the standardized rank sum statistic W is

$$z = \frac{W - \mu_W}{\sigma_W} = \frac{67.5 - 45.5}{6.745} = 3.26$$

Since we would expect W to have large values if the alternative were true, the approximate P-value is $P(Z \geq 3.26) = 0.0006$. There is very strong evidence that the supplemented birds miss the peak by more days than the control birds.

Exercise 25.25

(a) The null and alternative hypotheses are

H_0: median = 0

H_a: median > 0

The ranks of the absolute values are

Absolute values	Ranks
0.4	1.5
0.4	1.5
0.7	3
1.1	4
1.2	5
1.3	6
2.0	7.5
2.0	7.5
2.2	9
2.3	10

The Wilcoxon signed rank statistic is

$$W^+ = 1.5 + 3 + 4 + 5 + 7.5 + 7.5 + 9 + 10 = 47.5$$

The values for the mean and variance are

$$\mu_{W^+} = \frac{n(n+1)}{4} = \frac{10(11)}{4} = 27.5$$

and

$$\sigma_{W^+} = \sqrt{\frac{n(n+1)(2n+1)}{24}} = \sqrt{\frac{(10)(11)(21)}{24}} = 9.811$$

and the standardized signed rank statistic W^+ is

$$\frac{W^+ - \mu_{W^+}}{\sigma_{W^+}} \geq \frac{47.5 - 27.5}{9.811} = 2.04$$

If the cola lost sweetness, we would expect the differences (before storage – after storage) to be positive. Thus, the ranks of the positive observations should be large and we would expect the value of the statistic W^+ to be large when the alternative hypothesis is true. The approximate *P*-value is $P(Z \geq 2.04) = 0.021$. We conclude that the cola does lose sweetness in storage.

The output from the Minitab computer package on the next page gives a similar result. Minitab includes a correction to the standard deviation in the Normal approximation to account for the ties in the ranks, so. the *P*-value given by Minitab is slightly different than the one we obtained.

Wilcoxon Signed Rank Test

```
TEST OF MEDIAN = 0.000000 VERSUS MEDIAN G.T.  0.000000

              N FOR    WILCOXON              ESTIMATED
         N    TEST    STATISTIC   P-VALUE     MEDIAN
Loss    10     10        47.5     0.023       1.150
```

(b) The conclusions are the same and the P-values are also quite similar. The one-sample t test hypotheses are

$$H_0: \mu = 0$$

$$H_a: \mu > 0$$

Both tests assume that the tasters in the study are a simple random sample of all tasters. The one-sample t test also assumes that the (before storage) – (after storage) sweetness differences are Normally distributed.

Exercise 25.48

The null and alternative hypotheses for the Kruskal-Wallis test are

H_0: seedling growths have the same distribution in all groups

H_a: seedling growths are systematically higher in some groups than in others

When the distributions have the same shape, the null hypothesis for the Kruskal-Wallis is that the median growth in all groups are equal, and the alternative hypothesis is that not all four medians are equal.

The computations required for the Kruskal-Wallis test statistic follow:

```
Growth    3.2      4.6      5.0      5.3      5.4      5.8      7.4
Group   10000     5000     5000    10000     5000    10000     5000
Rank        1        2        3        4        5        6        7

Growth    7.5      8.2      9.1      9.2     10.8     11.1     11.1
Group   10000     1000        0        0        0     1000     1000
Rank        8        9       10       11       12     13.5     13.5

Growth   11.3     13.5
Group    1000        0
Rank       15       16
```

Nematodes	Ranks	Sum of ranks
0	10, 11, 12, 16	49
1000	9, 13.5, 13.5, 15	51
5000	2, 3, 5, 7	17
10000	1, 4, 6, 8	19

$$H = \frac{12}{N(N+1)} \sum \frac{R_i^2}{n_i} - 3(N+1) = \frac{12}{16(16+1)} \left(\frac{49^2}{4} + \frac{51^2}{4} + \frac{17^2}{4} + \frac{19^2}{4} \right) - 3(16+1) = 11.34$$

Since $I = 4$ groups, the sampling distribution of H is approximately chi-square with $4 - 1 = 3$ degrees of freedom. From Table E we see that the P-value is approximately 0.01. There is strong evidence of a difference in seedling growth between the four groups.

The MINITAB software gives the following output when doing the Kruskal-Wallis test. The medians, average ranks (in place of sums of ranks), H statistic and P-value are given. The H statistic with an adjustment for ties in the ranks is also given.

Kruskal-Wallis Test

```
LEVEL     NOBS     MEDIAN  AVE. RANK
  1         4      10.000      12.3
  2         4      11.100      12.8
  3         4       5.200       4.2
  4         4       5.550       4.7
OVERALL    16                   8.5

H = 11.34  d.f. = 3  p = 0.010
H = 11.35  d.f. = 3  p = 0.010 (adjusted for ties)
```

CHAPTER 26

STATISTICAL PROCESS CONTROL

OVERVIEW

In practice, work is often organized into a chain of activities that lead to some result. A chain of activities that turns inputs into outputs is called a **process**. A process can be described by a **flowchart**, which is a picture of the stages of a process. A **cause-and-effect diagram**, which displays the logical relationships between the inputs and output of a process, is also useful for describing and understanding a process.

All processes have variation. If the pattern of variation is stable over time, the process is said to be in statistical control. In this case, the sources of variation are called **common causes.** If the pattern is disrupted by some unusual event, **special cause** variation is added to the common cause variation. **Control charts** are statistical plots intended to warn when a process is disrupted or **out of control.**

Standard **3σ control charts** plot the values of some statistic Q for regular samples from the process against the time order in which the samples were collected. The **center line** of the chart is at the mean of Q. The **control limits** lie three standard deviations of Q above (the **upper control limit**) and below (the **lower control limit**) the center line. A point outside the control limits is an **out-of-control signal.** For **process monitoring** of a process that has been in control, the mean and standard deviations used to establish the center line and control limits are based on past data and are updated regularly.

When we measure some quantitative characteristic of a process, we use $\bar{x}$ and s **charts** for process control. The $\bar{x}$ chart plots the sample means of samples of size n from the process and the s chart the sample standard deviations. The s chart monitors variation within individual samples from the process. If the s chart is in control, the $\bar{x}$ chart monitors variation from sample to sample. To interpret charts, always look first at the s chart.

For a process that is in control with mean μ and standard deviation σ, the 3σ $\bar{x}$ chart based on samples of size n has center line and control limits

$$CL = \mu \quad UCL = \mu + 3\frac{\sigma}{\sqrt{n}} \quad LCL = \mu - 3\frac{\sigma}{\sqrt{n}}$$

The $3\sigma s$ chart has control limits

$$UCL = (c_4 + 2c_5)\sigma = B_6\sigma \quad LCL = (c_4 - 2c_5)\sigma = B_5\sigma$$

and the values of c_4, c_5, B_5, and B_6 can be found in Table 24.3 in your textbook for n from 2 to 10.

An **R chart** based on the range of observations in a sample is often used in place of an s chart. We will rely on software to produce these charts. Formulas can be found in books on quality control. $\bar{x}$ and R charts are interpreted the same way as $\bar{x}$ and s charts.

It is common to use various **out-of-control signals** in addition to "one point outside the control limits." In particular, a **runs signal** (nine consecutive points above the center line or nine consecutive points below the center line) for an $\bar{x}$ chart allows one to respond more quickly to a gradual drift in the process center.

We almost never know the mean μ and standard deviation σ of a process. They must be estimated from past data. We estimate μ by the mean $\bar{\bar{x}}$ of the observed sample means $\bar{x}$. We estimate σ by

$$\hat{\sigma} = \frac{\bar{s}}{c_4}$$

where $\bar{s}$ is the mean of the observed sample standard deviations. **Control charts based on past data** are used at the **chart setup** stage for a process that may not be in control. Start with control limits calculated from the same past data that you are plotting. Beginning with the s chart, narrow the limits as you find special causes and remove the points influenced by these causes. When the remaining points are in control, use the resulting limits to monitor the process.

Statistical process control maintains quality more economically than inspecting the final output of a process. Samples that are **rational subgroups** (subgroups that capture the features of the process in which we are interested) are important to effective control charts. A process in control is stable, so we can predict its behavior. If individual measurements have a normal distribution, we can give the **natural tolerances.**

A process is **capable** if it can meet or exceed the requirements placed on it. Control (stability over time) does not in itself improve capability. Remember that control describes the internal state of the process, whereas capability relates the state of the process to external specifications.

There are control charts for several different types of process measurements. One important type is the **p chart,** a control chart based on plotting sample proportions $\hat{p}$ from regular samples from a process against the order in which the samples were taken. We estimate the process proportion p of "successes" by

$$\bar{p} = \frac{\text{total number of successes in past samples}}{\text{total number of opportunities in these samples}}$$

and then the control limits for a p chart for future samples of size n are

$$\text{UCL} = \bar{p} + 3\sqrt{\frac{\bar{p}(1-\bar{p})}{n}} \quad \text{CL} = \bar{p} \quad \text{LCL} = \bar{p} - 3\sqrt{\frac{\bar{p}(1-\bar{p})}{n}}$$

The interpretation of p charts is very similar to that of $\bar{x}$ charts. The out-of-control signals used are also the same as for $\bar{x}$ charts.

GUIDED SOLUTIONS

Exercise 26.1

KEY CONCEPTS: Flowcharts and cause-and-effect diagrams

For this exercise, it is important to choose a process that you know well so that you can describe it carefully and recognize those factors that affect the process. Use the space provided for your flowchart and cause-and-effect diagram.

Exercise 26.4

KEY CONCEPTS: Pareto charts

What percent of total losses do these 9 DRGs account for?

Sum of percent losses =

Use the axes to make your Pareto chart.

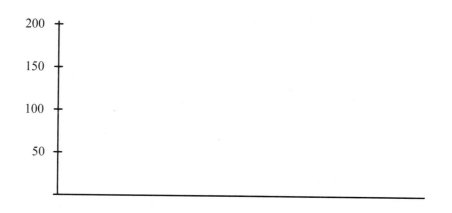

DRG

Which DRGs should the hospital study first when attempting to reduce its losses?

Exercise 26.7

KEY CONCEPTS: Common causes

Refer to Exercise 27.1 in this Study Guide. For a process you know well, what are some common sources of variation in the process?

What are some special causes that might drive the process out of control?

Exercise 26.15

KEY CONCEPTS: $\bar{x}$ and s charts

For the first two samples in Figure 27.10 of your textbook compute $\bar{x}$ and s.

Sample 1

$\bar{x} =$

$s =$

Sample 2

$\bar{x} =$

$s =$

If you have access to statistical software, use the software to make your $\bar{x}$ and s charts. Otherwise, to make the $\bar{x}$ chart, compute

$$UCL = \mu + 3\frac{\sigma}{\sqrt{n}} =$$

$$CL = \mu =$$

$$LCL = \mu - 3\frac{\sigma}{\sqrt{n}} =$$

Plot the UCL, CL, LCL, and values of $\bar{x}$ for all 18 samples.

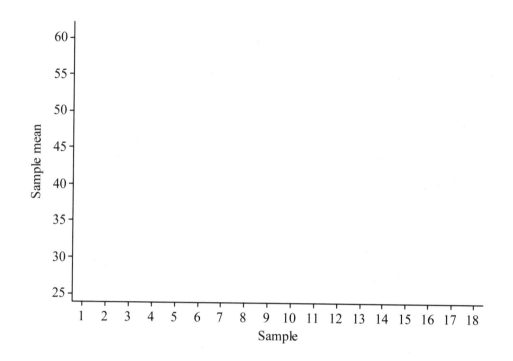

To make the s chart, compute

$$\text{UCL} = B_6\sigma =$$

$$\text{CL} = c_4\sigma =$$

$$\text{LCL} = B_5\sigma =$$

Plot the UCL, CL, LCL, and the values of s for all 18 samples in the chart that follows. How would you describe the state of the process?

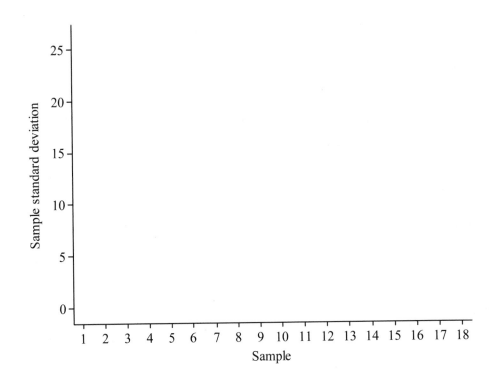

Exercise 26.20

KEY CONCEPTS: $\bar{x}$ and s control charts using past data

(a) From the values of $\bar{x}$ and s in Table 27.1 of your textbook, compute (by hand, a calculator, or using software)

$\bar{\bar{x}}$ = mean of the 20 values of $\bar{x}$ =

$\bar{s}$ = mean of the 20 values of s =

Hence we estimate μ to be

$\hat{\mu} = \bar{\bar{x}}$ =

and we estimate σ to be

$$\hat{\sigma} = \frac{\bar{s}}{c_4} =$$

(b) Look at the s chart in Figure 27.7 of your textbook. What patterns do you see that might suggest that the process σ may now be less than 43 mV?

Exercise 26.29

KEY CONCEPTS: Natural tolerances

The natural tolerances are $\mu \pm 3\sigma$. We do not know μ and σ, so we must estimate them from the data. We remove sample 5 from the data. Based on the remaining 17 samples, estimate

$\bar{\bar{x}}$ = mean of the 17 values of $\bar{x}$ =

$\bar{s}$ = mean of the 17 values of s =

Hence, we estimate μ to be

$$\hat{\mu} = \bar{\bar{x}} =$$

and we estimate σ to be

$$\hat{\sigma} = \frac{\bar{s}}{c_4} =$$

Based on these estimates, the natural tolerances for the distance between the holes are

$$\hat{\mu} \pm 3\hat{\sigma} =$$

Exercise 26.30

KEY CONCEPTS: Capability

Refer to Exercise 24.29 in this Study Guide. Based on the 17 samples that were in control, we see that estimates of μ and σ are $\hat{\mu} = 43.41$ and $\hat{\sigma} = 12.39$. We therefore assume that distances between holes vary from meter to meter according to an $N(43.41, 12.39)$ distribution. Use Normal probability calculations to find the probability that the distance x between holes in a randomly selected meter is between 54 ± 10 (i.e., between 44 and 64). Refer to Chapter 3 of your textbook if you have forgotten how to do normal probability calculations.

$$P(44 < x < 64) =$$

We conclude that about what percent of meters meet specifications?

Exercise 26.34

KEY CONCEPTS: p charts

To find the appropriate center line and control limits, we must first compute $\bar{p}$. The total number of opportunities for missing or deformed rivets is just the total number of rivets because each rivet has the possibility of being missing or deformed. The number of "successes" in past samples is just the missing or deformed rivets in the recent data. What are these values? Now, estimate $\bar{p}$:

$$\bar{p} = \frac{\text{total number of successes in past samples}}{\text{total number of opportunities in these samples}} =$$

The next wing contains $n = 1070$ rivets, and the control limits for a p chart for future samples of size $n = 1070$ are

$$\text{UCL} = \bar{p} + 3\sqrt{\frac{\bar{p}(1-\bar{p})}{n}} =$$

$$\text{CL} = \bar{p} =$$

$$\text{LCL} = \bar{p} - 3\sqrt{\frac{\bar{p}(1-\bar{p})}{n}} =$$

COMPLETE SOLUTIONS

Exercise 26.1

We take as our example the process of making a cup of coffee. A possible flowchart and cause-and-effect diagram of the process follow:

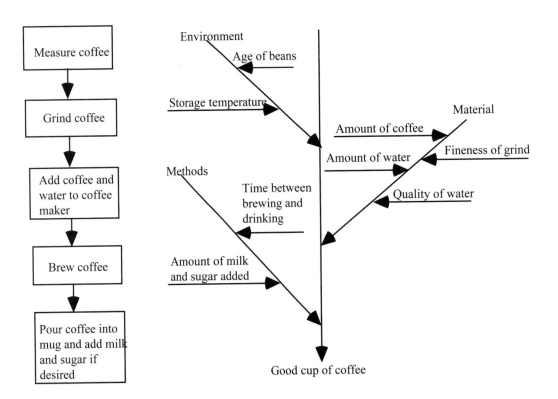

Exercise 26.4

Adding the percents listed, the percent of total losses that these 9 DRGs account for is 80.5%. A Pareto chart of losses by DRG follows:.

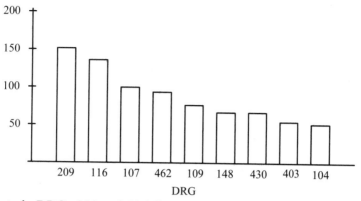

The hospital ought to study DRGs 209 and 116 first in attempting to reduce its losses. These are the two DRGs with the largest percent losses and combined account for nearly 30% of all losses.

Exercise 26.7

In Exercise 27.1 of this Study Guide, we described the process of making a good cup of coffee. Some sources of common-cause variation are variation in how long the coffee has been stored and the conditions under which it has been stored, variation in the measured amount of coffee used, variation in how finely the coffee is ground, variation in the amount of water added to the coffee maker, variation in the length of time the coffee sits between when it has finished brewing and when it is drunk, and variation in the amount of milk and/or sugar added.

Some special causes that might at times drive the process out of control would be a bad batch of coffee beans, a serious mismeasurement of the amount of coffee used or the amount of water used, a malfunction of the coffee maker or a power outage, interruptions that result in the coffee sitting a long time before it is drunk, and the use of milk that has gone bad.

Exercise 26.15

We compute $\bar{x}$ and s for the first two samples:

First sample: $\bar{x} = 48, s = 8.94$ Second sample: $\bar{x} = 46, s = 13.03$

To make the $\bar{x}$ chart, we note that

$$\text{UCL} = \mu + 3\frac{\sigma}{\sqrt{n}} = 43 + 3\frac{12.74}{\sqrt{5}} = 43 + 17.09 = 60.09$$

$$\text{CL} = \mu = 43$$

$$\text{LCL} = \mu - 3\frac{\sigma}{\sqrt{n}} = 43 - 3\frac{12.74}{\sqrt{5}} = 43 - 17.09 = 25.91$$

resulting in the chart that follows:

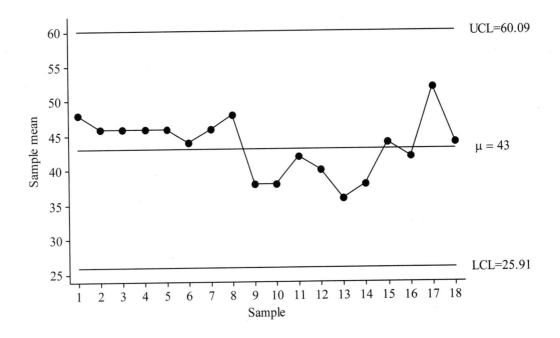

To make the *s* chart, we note that

$$UCL = B_6\sigma = 1.964(12.74) = 25.02$$

$$CL = c_4\sigma = 0.9400(12.74)) = 11.98$$

$$LCL = B_5\sigma = 0(12.74)) = 0$$

resulting in the chart that follows.

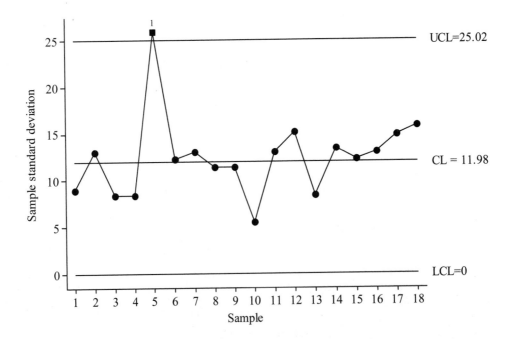

The s chart shows a lack of control at sample point 5, but otherwise neither chart shows a lack of control. We would want to find out what happened at sample 5 to cause a lack of control in the s chart.

Exercise 26.20

(a) From the values of $\bar{x}$ and s in Table 27.1 of the textbook, we compute (using software)

$$\bar{\bar{x}} = \text{mean of the 20 values of } \bar{x} = 275.065$$
$$\bar{s} = \text{mean of the 20 values of } s = 34.55$$

Hence, we estimate μ to be

$$\hat{\mu} = \bar{\bar{x}} = 275.065$$

and we estimate σ to be (using the fact that the samples are each of size $n = 4$ and according to Table 24.3 of the textbook, $c_4 = 0.9213$)

$$\hat{\sigma} = \frac{\bar{s}}{c_4} = \frac{34.55}{0.9213} = 37.5$$

(b) If we look at the s chart in Figure 27.7 of the textbook we see that most of the points lie below 40 (and more than half of the points below 40 lie well below 40), while of the points above 40, all but one (sample 12) are only slightly larger than 40. The s chart suggests that typical values of s are below 40, which is consistent with the estimate of σ in part (a).

Exercise 26.29

The natural tolerances are $\mu \pm 3\sigma$. We do not know μ and σ, so we must estimate them from the data. We remove sample 5 from the data. Based on the remaining 17 samples, we find

$$\bar{\bar{x}} = \text{mean of the 17 values of } \bar{x} = 43.41$$
$$\bar{s} = \text{mean of the 17 values of } s = 11.65$$

Hence, we estimate μ to be

$$\hat{\mu} = \bar{\bar{x}} = 43.41$$

and we estimate σ to be (using the fact that the samples are each of size $n = 5$ and according to Table 24.3 of the textbook, $c_4 = 0.9400$)

$$\hat{\sigma} = \frac{\bar{s}}{c_4} = \frac{11.65}{0.9400} = 12.39$$

Based on these estimates, the natural tolerances for the distance between the holes are

$$\hat{\mu} \pm 3\hat{\sigma} = 43.41 \pm 3(12.39) = 43.41 \pm 37.17 \text{ or } 6.24 \text{ to } 80.58$$

Exercise 26.30

Based on the 17 samples that were in control, we saw in Exercise 27.29 in this Study Guide that estimates of μ and σ are $\hat{\mu} = 43.41$ and $\hat{\sigma} = 12.39$. We therefore assume that distances between holes vary from meter to meter according to an $N(43.41, 12.39)$ distribution. The probability that the distance x between holes in a randomly selected meter is between 54 ± 10 (i.e., between 44 and 64) is thus

$$P(44 < x < 64) = P(\frac{44 - 43.41}{12.39} < \frac{x - 43.41}{12.39} < \frac{64 - 43.41}{12.39}) = P(0.05 < Z < 1.66)$$

$$= P(Z < 1.66) - P(Z < 0.05) = 0.9515 - 0.5199 = 0.4316$$

We conclude that about 43.16% of meters meet specifications.

Exercise 26.34

The total number of opportunities for missing or deformed rivets is just the total number of 34,700 rivets because each rivet has the possibility of being missing or deformed. The number of "successes" in past samples is just the 208 missing or deformed rivets in the recent data. We therefore estimate the process proportion p of "successes" from the recent data by

$$\bar{p} = \frac{\text{total number of successes in past samples}}{\text{total number of opportunities in these samples}} = \frac{208}{34,700} = 0.00599$$

The next wing contains $n = 1070$ rivets, and the control limits for a p chart for future samples of size $n = 1070$ are

$$\text{UCL} = \bar{p} + 3\sqrt{\frac{\bar{p}(1-\bar{p})}{n}} = 0.00599 + 3\sqrt{\frac{0.00599(1-0.00599)}{1070}} = 0.00599 + 0.00708 = 0.01307$$

$$\text{CL} = \bar{p} = 0.00599$$

$$\text{LCL} = \bar{p} - 3\sqrt{\frac{\bar{p}(1-\bar{p})}{n}} = 0.00599 - 3\sqrt{\frac{0.00599(1-0.00599)}{1070}} = 0.00599 - 0.00708 = 0$$

Note that in the LCL, we set negative values to 0 because a proportion can never be less than 0.

CHAPTER 27

MULTIPLE REGRESSION

OVERVIEW

Multiple linear regression extends the techniques of simple linear regression to situations involving $p > 1$ explanatory variables $x_1, x_2, \ldots, x_p$. The data consist of the values of the response y and the p explanatory variables for n individuals or cases. Data analysis begins by examining the distribution of the variables individually and then drawing scatterplots to explore the relationships between the variables.

The mean response μ_y for a **multiple regression model** based on p explanatory variables $x_1, x_2, \ldots, x_p$ is

$$\mu_y = \beta_0 + \beta_1 x_1 + \beta_2 x_2 + \ldots + \beta_p x_p$$

The multiple regression equation predicts the value of the response y as a linear function of the explanatory variables

$$\hat{y}_i = b_0 + b_1 x_{i1} + b_2 x_{i2} + \ldots + b_p x_{ip}$$

where the coefficients b_i are estimated using the method of least squares. The variability of the responses about the multiple regression equation is measured in terms of the **regression standard error** s,

$$s = \frac{\sum e_i^2}{n - p - 1}$$

where the e_i are the **residuals**:

$$e_i = y_i - \hat{y}_i$$

The regression standard error s has $n - p - 1$ degrees of freedom. The **distribution of the residuals** should be examined and the residuals should be plotted against each of the p explanatory variables. In practice, the b's and s are calculated using statistical software.

A special case of the multiple linear regression model is fitting separate regression lines to two sets of data. Fitting the lines is done using an **indicator variable** to show from which data set an observation comes and using an **interaction** term to allow for different slopes.

The **ANOVA table** for a multiple regression is analogous to that in simple linear regression. It gives the sum of squares, the mean squares, and the degrees of freedom for regression and residual sources of variation. The ANOVA F is the regression mean square (MSM) divided by the residual mean square (MSE) and is used to test the hypothesis $H_0: \beta_1 = \beta_2 = \ldots = \beta_p = 0$. Under H_0, this statistic has an $F(p, n - p - 1)$ distribution.

The **squared multiple correlation** can be written as the ratio of model to total variation, namely,

$$R^2 = \text{SSM/SST}$$

and is interpreted as the proportion of the variability in the response variable y that is explained by the explanatory variables $x_1, x_2, \ldots, x_p$ in the multiple regression.

A **level C confidence interval** for β_j is

$$b_j \pm t^* \, \text{SE}_{b_j}$$

where t^* is the upper $(1 - C)/2$ critical value for the $t(n - p - 1)$ distribution. SE_{b_j} is the standard error of b_j and in practice is computed using statistical software.

The **test of the hypothesis** $H_0: \beta_j = 0$ is based on the *t* **statistic**

$$t = \frac{b_j}{\text{SE}_{b_j}}$$

with P-values computed from the $t(n - p - 1)$ distribution. In practice, statistical software is used to carry out these tests.

In multiple regression, interpretation of these confidence intervals and tests depends on the particular explanatory variables in the multiple regression model. The estimate of β_j represents the effect of the explanatory variable x_j when it is added to a model already containing the other explanatory variables. The test of $H_0: \beta_j = 0$ tells us if the improvement in the ability of our model to predict the response y by adding x_j to a model already containing the other explanatory variables is statistically significant. It does not tell us if x_j would be useful for predicting the response in multiple regression models with a different collection of explanatory variables.

Confidence intervals for the mean response μ_y have the form

$$\hat{y} \pm t^* \, \text{SE}_{\hat{\mu}}$$

Prediction intervals for an individual future response y have the form

$$\hat{y} \pm t^* \, \text{SE}_{\hat{y}}$$

where t^* is the critical value for the $t(n - p - 1)$ density curve. $\text{SE}_{\hat{\mu}}$ and $\text{SE}_{\hat{y}}$ can be computed using statistical software. In practice, both confidence intervals for μ_y and prediction intervals for an individual future observation are computed using statistical software.

GUIDED SOLUTIONS

Exercise 27.15

KEY CONCEPTS: Regression with indicator variables

(a) Review Exercise 4.7 if you have forgotten how to make a scatterplot using separate symbols for a categorical variable. If you are not using software to make the plot, use the axis that follow for your plot. To get you started, we have plotted the first point for men (using the symbol x) and the first point for women (using the symbol o).

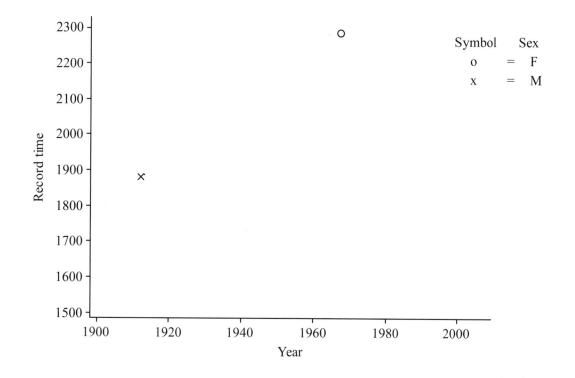

How would you describe the pattern for each sex? Do the points for each sex tend to follow a straight line or some curved relation?

How would you describe the progress of men and women?

(b) To fit a model with two regression lines, one for men and one for women, you will need to create an indicator variable for sex. To allow for lines of different slopes, you will also need to create a variable representing the interaction between sex and year. After doing so, use software to fit a multiple regression model with year, the indicator variable for sex, and the variable representing the interaction between sex and year as predictors.

Estimated model with two regression lines:

Estimated regression line for men:

Estimated regression line for women:

(c) Do the data appear to support any of these claims? If you know recent world record times for men and women, you might see if the rate of improvement for women has changed and if the difference in record times for men and women has become negligible.

Exercise 27.25

KEY CONCEPTS: Multiple linear regression, R^2, F test, t tests

The following Minitab output for the regression of weight on length and width can be used to help answer parts (a) through (d). You should run the regression with the software you are using in your course to become familiar with the format of the output. Although the regression output should be similar to the Minitab output, there may be slight variations in the names for some of the quantities.

Regression Analysis: Weight versus Length, Width

```
The regression equation is
Weight = - 579 + 14.3 Length + 113 Width

Predictor      Coef   SE Coef      T       P
Constant    -578.76     43.67  -13.25   0.000
Length       14.307     5.659    2.53   0.014
Width       113.50      30.26    3.75   0.000

S = 88.6760   R-Sq = 93.7%   R-Sq(adj) = 93.5%
```

```
Analysis of Variance

Source          DF      SS        MS       F       P
Regression       2   6229332   3114666   396.09   0.000
Residual Error  53    416762      7863
Total           55   6646094
```

(a) As part of the Minitab output, the formula for the estimated regression equation is provided. If you are using a different software package, you may need to use the estimated coefficients to write the equation. Use the information in the output to give the estimated multiple regression equation

$\hat{y} =$

(b) Which regression quantity measures the amount of variation in the response explained by the model? It is included in the output.

Amount of variation in weight explained by the model in (a) =

(c) The null and alternative hypotheses tested by the ANOVA F test are

H_0:

H_a:

Does a test of these hypotheses answer the question posed? Both the test statistic and P-value are included with the output.

(d) The individual t tests that β_1 and β_2 are significantly different from zero are included in the output. What do you conclude from them?

The following Minitab output for the regression of weight on length and width and their interaction can be used to help answer parts (e) through (h). When using your software, you will need to first create a new column for the product of length and width and then include this variable "Interaction" in the model.

Regression Analysis: Weight versus Length, Width, Interaction

```
The regression equation is
Weight = 114 - 3.48 Length - 94.6 Width + 5.24 Interaction


Predictor        Coef   SE Coef      T      P
Constant       113.93     58.78   1.94  0.058
Length         -3.483     3.152  -1.10  0.274
Width          -94.63     22.30  -4.24  0.000
Interaction    5.2412    0.4131  12.69  0.000


S = 44.2381    R-Sq = 98.5%    R-Sq(adj) = 98.4%


Analysis of Variance

Source         DF        SS       MS       F       P
Regression      3   6544330  2181443  1114.68  0.000
Residual Error 52    101765     1957
Total          55   6646094
```

(e) As part of the Minitab output, the formula for the estimated regression equation is provided. If you are using a different software package, you may need to use the estimated coefficients to write the equation. Use the information in the output to give the estimated multiple regression equation.

$\hat{y} =$

(f) Which regression quantity measures the amount of variation in the response explained by the model? It is included in the output.

Amount of variation in weight explained by the model in (e) =

(g) The null and alternative hypotheses tested by the ANOVA F test are

H_0:

H_a:

Does a test of these hypotheses answer the question posed? Both the test statistic and P-value are included with the output.

(h) When the explanatory variables are correlated, the estimated coefficients change as well as their individual t statistics. Since the interaction term is the product of length and width, it is correlated with both length and width. Describe how the individual t statistics change when the interaction term is added.

Exercise 27.27

KEY CONCEPTS: Confidence intervals for the mean, prediction intervals

Confidence intervals for the mean and prediction intervals require specifying a list of values for all the explanatory variables in the model. You are asked to obtain these intervals for the tenth perch. What are the values of the explanatory variables for this perch?

Length =

Width =

Interaction =

Software packages differ in how they obtain confidence and prediction intervals. In some packages, such as SAS, if you ask for these intervals they are automatically produced for the explanatory variables at every observation. In other packages, such as Minitab, you must specify the explanatory variables for which you want confidence and prediction intervals. You should learn how to obtain these intervals with the software you are using for this course. The Minitab output follows:

```
Predicted Values for New Observations

New
Obs     Fit  SE Fit        95% CI             95% PI
  1   84.02   10.41  (63.13, 104.91)  (-7.18, 175.21)


Values of Predictors for New Observations

New
Obs  Length  Width  Interaction
  1    21.0   2.80         58.8
```

What t distribution was used to obtain these intervals?

Interpret both intervals.

Exercise 27.29

KEY CONCEPTS: Residual plots

Recall that the conditions for inference require agreement between the observed and predicted values (residuals centered about a horizontal line through 0), constant variance (the residuals look like an unstructured band of points centered around a horizontal line through 0), and Normality (absence of outliers in the residual plot).

Do you see any problems in either of the plots?

COMPLETE SOLUTIONS

Exercise 27.15
(a)

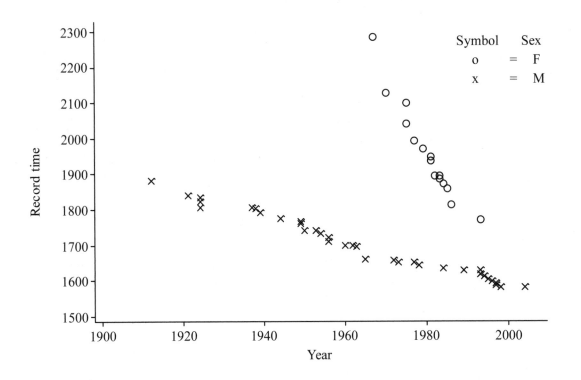

For both men and women the points tend to follow a straight line. Thus, we would describe the pattern as linear and decreasing.

The progress for women appears better than for men. Times for women have been decreasing more rapidly than the times for men.

(b) We used the variable "ind" as the indicator variable for sex. Ind = 1 represents males and ind = 0 females. The interaction term is the variable ind*year. Here are the Minitab results of the fitted lines:

```
The regression equation is
Record time  = 41373 - 19.9 year - 33247 ind + 16.6 ind*year


Predictor        Coef  SE Coef        T       P
Constant        41373     1713    24.15   0.000
year         -19.9046   0.8652   -23.01   0.000
ind            -33247     1734   -19.17   0.000
ind*year      16.6326   0.8759    18.99   0.000


S = 21.1736   R-Sq = 98.3%   R-Sq(adj) = 98.2%
```

Based on these results we have

Estimated model with two regression lines: record time = 41373 – 19.9 × year – 33247 × ind + 16.6 × ind*year

Estimated regression line for men: record time = (41373 – 33247) + (–19.9 + 16.6) × year
$$= 8126 - 3.3 \times \text{year}$$

Estimated regression line for women: record time = 41373 – 19.9 × year

(c) The improvement for women (at least for the data we have) is more rapid than for men. The difference in record times is decreasing, but it is not negligible. One is tempted to extrapolate the data and argue that the difference will become negligible as time passes. However, extrapolation is dangerous. If one looks at the trend in the two lines, one might expect that the difference in world record times for men and women to be negligible by about the year 2000 and perhaps that women would have a faster world record time than men after 2000. This is not the case, at least as of 2006.

Exercise 27.25

(a) The estimated multiple regression equation can be read off the Minitab output as

```
    Weight = - 579 + 14.3 Length + 113 Width
```

Alternatively, the estimated equation can be obtained from the coefficient column in the output reading the values for the constant, length coefficient, and width coefficient to yield

$$\hat{y} = -578.76 + 14.307 \text{ length} + 113.50 \text{ width}$$

(b) The quantity R^2 measures the variability explained by the model, or in this case the variation in the weight of a perch explained by its length and width. R^2 is given on the Minitab output as

```
                R-Sq = 93.7%
```

Thus, almost 94% of the variation is the weight of a perch is explained by its length and width.

(c) The null and alternative hypotheses tested by the ANOVA F test are

$$H_0: \beta_1 = \beta_2 = 0$$

$$H_a: \text{at least one of } \beta_1 \text{ and } \beta_2 \text{ is not } 0$$

where β_1 is the regression coefficient of "length" and β_2 is the regression coefficient of "width" in our multiple linear regression model. From the output, $F = 396.09$ and the P-value $= 0.000$. We reject the null hypothesis, so at least one of length or width is helpful in predicting the weight of the perch.

(d) The individual t statistic for length is $t = 2.53$ with P-value $= 0.014$. This indicates that length helps explain the weight of the perch even after we allow width to explain the weight. The individual t statistic for length is $t = 3.75$ with P-value $= 0.000$. This indicates that width helps explain the weight of the perch, even after we allow length to explain the weight. The combined information from these indicates that both variables should be used to explain the weight of the perch.

(e) The estimated multiple regression equation can be read off the Minitab output as

```
Weight = 114 - 3.48 Length - 94.6 Width + 5.24 Interaction
```

Alternatively, the estimated equation can be obtained from the coefficient column for the constant, length coefficient, width coefficient, and interaction coefficient to yield

$$\hat{y} = 114 - 3.483 \text{ length} - 94.63 \text{ width} + 5.2412 \text{ interaction}$$

(f) The quantity R^2 measures the variability explained by the model, or in this case the variation in the weight of a perch explained by its length and width. R^2 is given on the Minitab output as

```
R-Sq = 98.5%
```

Thus almost 99% of the variation is the weight of a perch is explained by its length, width, and their interaction.

(g) The null and alternative hypotheses tested by the ANOVA F test are

$$H_0: \beta_1 = \beta_2 = \beta_3 = 0$$

$$H_a: \text{at least one of } \beta_1, \beta_2 \text{ and } \beta_3 \text{ is not } 0$$

where β_1 is the regression coefficient of "length," β_2 is the regression coefficient of "width," and β_3 is the regression coefficient of interaction in our multiple linear regression model. From the output, $F = 1114.68$ and the P-value $= 0.000$. We reject the null hypothesis, so at least one of length, width, or their interaction is helpful in predicting the weight of the perch.

(h) When interaction is added, the individual t for length changes from 2.53 to -1.10 and the P-value changes from 0.014 to 0.274. Not only is the statistical significance affected, but the sign of the estimated coefficient changes from positive to negative. The individual t for width changes from 3.75 to –4.24, with the statistical significance unaffected. However, the sign of the estimated coefficient has changed. The meaning of the coefficients for length and width change with the addition of the interaction term, so these changes are not surprising. In this example, we can see that the relationship between the response y and any one explanatory variable can change greatly depending on what other explanatory variables are present in the model.

Exercise 27.27

A $t(n - p - 1) = t(56 - 3 - 1)$, or t distribution with 52 degrees of freedom, was used to obtain the intervals. We are 95% confident that the average weight of all fish with a length of 21 and a width of 2.8 is between 63.13 and 104.91. We are 95% confident that a new fish caught with this length and width will have a weight no larger that 175.21. We can take the lower endpoint as zero since negative weights are not possible.

Exercise 27.29

The first plot has somewhat of a funnel shape, with variability increasing as the variable Purchase12 increases. Otherwise, there are no other obvious problems (no outliers and the residuals appear to be centered about 0).

The second plot also has a funnel shape with the variability decreasing as Recency increases. The points appear to be centered about 0 and there are no outliers in the vertical direction, but the point at the far right is an outlier in the horizontal directions. This may be an influential observation.

We conclude that the condition that appears to be most suspect is the constant variance assumption. There is nothing in either plot that suggests there are serious problems with the assumption of Normality or independence. We might be concerned about the effect of the possible influential observation on the fitted model (especially the effect on the impact of the predictor variable Recency).

CHAPTER 28

MORE ABOUT ANALYSIS OF VARIANCE

OVERVIEW

Two-way analysis of variance is designed to compare the means of populations that are classified according to two factors R and C in a **two-way layout.** As with one-way ANOVA, the populations are assumed to be normal with possibly different means and the same standard deviation. The observations are independent SRSs drawn from each population.

Typically the means are summarized in a two-way table with the rows corresponding to the factor R and the columns corresponding to factor C. These means are should be plotted so that the **main effects** of each factor can be examined as well as their **interaction.**

The two-way ANOVA table organizes the calculations required to compute F-statistics and P-values to test three hypotheses: no main effect for factor R, no main effect for factor C, and no interaction between the factors.

Tukey pairwise multiple comparisons are often used as a follow up analysis in both one-way and two-way ANOVA. These multiple comparisons are designed to determine which pairs of population means are different and to give confidence intervals for the differences all pairs of means with an **overall level of confidence.**

GUIDED SOLUTIONS

Exercise 28.5

KEY CONCEPTS: Tukey's simultaneous confidence intervals in a one-way ANOVA

(a) How many comparisons are there when we compare four colors, or any four treatments.   In this exercise, you can count the number by making a list. There is a general formula in terms of the number of the treatments given in the Complete Solution.

(b) You should learn how to use your software to calculate Tukey's simultaneous confidence intervals for all pairwise comparisons of colors. The Minitab output is reproduced below to help you to complete this exercise. C1 refers to the column containing the color associated with each observation.

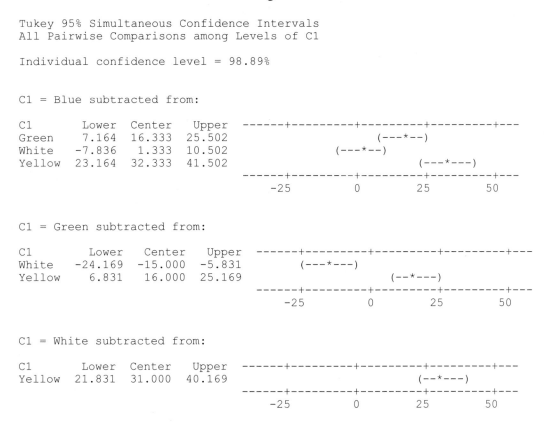

```
Tukey 95% Simultaneous Confidence Intervals
All Pairwise Comparisons among Levels of C1

Individual confidence level = 98.89%


C1 = Blue subtracted from:

C1        Lower   Center   Upper   ------+---------+---------+---------+---
Green     7.164   16.333   25.502                      (---*--)
White    -7.836    1.333   10.502                (---*--)
Yellow   23.164   32.333   41.502                              (---*---)
                                   ------+---------+---------+---------+---
                                       -25        0        25        50


C1 = Green subtracted from:

C1         Lower    Center    Upper   ------+---------+---------+---------+---
White    -24.169   -15.000   -5.831          (---*---)
Yellow     6.831    16.000   25.169                        (--*---)
                                      ------+---------+---------+---------+---
                                          -25        0        25        50


C1 = White subtracted from:

C1        Lower    Center   Upper   ------+---------+---------+---------+---
Yellow   21.831   31.000   40.169                          (--*---)
                                    ------+---------+---------+---------+---
                                        -25        0        25        50
```

Remember that we reject the null hypothesis that the means for a pair of populations are equal whenever the confidence interval doesn't include zero. In the output, the lower endpoint of the confidence interval is given below the word "lower,",the term "center" gives the difference in the two sample means (this is the center of the confidence interval), and the upper endpoint of the confidence interval is given below the word "upper."

Which pairs of colors are significantly different when we require significance level 5% for all the comparisons as a group?

Is yellow significantly better than every other color for attracting beetles?

Exercise 28.27

KEY CONCEPTS: Tukey's simultaneous confidence intervals in a one-way ANOVA

(a) When we use Tukey's simultaneous confidence intervals for the three differences in means, how confident are we that all three intervals capture the true differences?

(b) The one-way ANOVA gives strong evidence ($F = 259.12$, $P < 0.0001$) that the three population mean lengths are not equal. Now, use the results of the Tukey 95% confidence intervals to do a followup analysis. Your analysis should describe which pairs of population mean lengths are different as well as which populations have the larger mean lengths.

Exercise 28.33

KEY CONCEPTS: Four-step rule, two-way analysis of variance

The four-step process follows:

   State. What is the practical question that requires a statistical test?

   Plan. Identify the parameters, state null and alternative hypotheses, and choose appropriate test.

   Solve. (1) Check the conditions, (2) calculate the test statistic, and (3) find the *P*-value.

   Conclude. Return to the practical question to describe your results in this setting.

To apply the steps to this problem, here are some suggestions. You may want to use Examples 28.12, 28.13, and 28.14 of your text as a guide.

State. Using the language in the problem, state the goals of the study.

Plan. Be sure to check to make sure that you can safely use ANOVA.

Solve. The following output is from MINITAB. You should try and learn how to use your software to duplicate these calculations. We begin with means and standard deviations for each of the four groups and a boxplot of the four groups.

Descriptive Statistics: Count

Results for Sex = F

```
Variable  Housing    N     Mean   StDev
Count     Group     12    38.25   17.22
          Isolated  12    35.75   16.47
```

Results for Sex = M

```
Variable  Housing    N     Mean   StDev
Count     Group     12    25.25   13.42
          Isolated  12    17.00    8.21
```

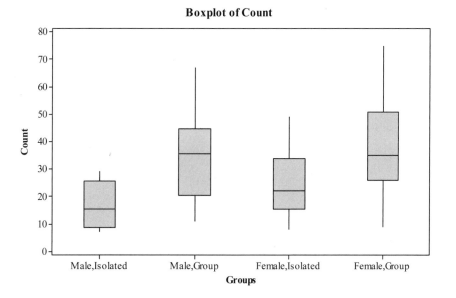

Boxplot of Count

Does it seem like the conditions for ANOVA are met? What is the ratio of the largest to the smallest standard deviation for the groups? Are the sample sizes large enough to apply ANOVA safely?

The interaction plot is given below. Make sure you understand how the four group means are entered in the plot.

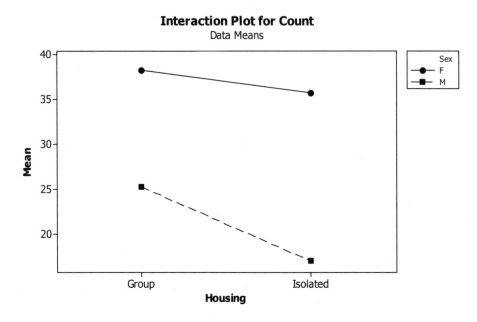

Try to identify some of the main features of the plot. Focus on the differences between sex, housing, and any interaction between them.

The ANOVA table follows:

Two-way ANOVA: Count versus Sex, Housing

```
Source        DF       SS        MS      F      P
Sex            1   3024.2   3024.19  14.84  0.000
Housing        1    346.7    346.69   1.70  0.199
Interaction    1     99.2     99.19   0.49  0.489
Error         44   8964.8    203.74
Total         47  12434.8

S = 14.27   R-Sq = 27.91%   R-Sq(adj) = 22.99%
```

Are the results of the significance tests in agreement with the impressions that you formed from the interaction plot?

Conclude. Using the plots and ANOVA table, summarize your results in words related to the problem.

COMPLETE SOLUTIONS

Exercise 28.5

KEY CONCEPTS: Tukey's simultaneous confidence intervals in a one-way ANOVA

(a) The six pairwise comparison are "Blue vs. Green," "Blue vs. White," "Blue vs. Yellow," "Green vs. White," "Green vs. Yellow" and "White vs. Yellow." When there are k treatments, there are $k(k - 1)/2$ pairwise comparisons. In this case, $k = 4$ giving $(4)(4 - 1)/2 = 6$ pairwise comparisons.

(b) Inspection of the confidence intervals shows that with the exception of "Blue vs. White," all other confidence intervals do not include zero. Thus, the pairs of colors that are significantly different from each other are "Blue vs. Green," "Blue vs. Yellow," "Green vs. White," "Green vs. Yellow" and "White vs. Yellow." Examining the intervals, we see that yellow is significantly better than every other color for attracting beetles.

Exercise 28.27

(a) Tukey's simultaneous confidence intervals provide an overall confidence level. That is, 95% refers to our level of confidence that *all three* of the intervals simultaneously capture the true pairwise differences.

(b) We interpret a confidence interval using Tukey's simultaneous confidence interval in the usual way. The first confidence interval for $\mu_{bihai} - \mu_{red}$ is 6.752 to 9.021. Since the interval does not include zero, there is a significant difference between the two means with the bihai variety having a larger mean length. The summary of the results follows.

The one-way ANOVA gives strong evidence ($F = 259.12$, $P < 0.0001$) that the three population mean lengths are not equal. None of the three Tukey's 95% simultaneous confidence intervals contain 0, which shows a significant difference between all pairs of means. The *bihai* variety has the largest mean length, being larger than both the red and yellow varieties. The red has a significantly larger length than the yellow variety. Although statistically significant, the confidence intervals show difference between the red and yellow varieties to be smaller than the difference between the *bihai* and both the red and yellow varieties.

Exercise 28.33

KEY CONCEPTS: Four-step rule, two-way analysis of variance

The four-step process follows:

State. Psychologists were interested in the effect of social isolation on behavior of hooded rats during a critical developmental period. They were also interested in whether this effect was the same for male and female rats. Twenty-four female rats were randomly assigned to either isolated or group housing, and the same was done for twenty-four male rats. Later the rats were observed in a group setting, and for each rat, the number of times each of three types of behavior (object play, locomotor play, and social play) occurred.

Plan. The sample means will be plotted and examined for interaction and main effects. The conditions for ANOVA inference will be checked. Two-way ANOVA *F* tests will be conducted to determine the significance of interaction and main effects. If necessary, Tukey pairwise comparisons will be used to identify differences among treatments.

Solve. The conditions for ANOVA seem satisfied. The ratio of the largest to the smallest standard deviation is $17.22/8.21 = 2.09$, which is slightly larger than 2. However, the sample sizes are all equal and of a moderate size, so the assumption of constant variance among the four treatment groups is not violated seriously enough to affect the conclusions. The smaller variability for the male isolated group is apparent in the boxplots. The boxplots also show distributions that are approximately symmetric with no high or low outliers. There do not appear to be any serious violations which would invalidate the use of two-way ANOVA *F* tests to analyze these data.

The ANOVA table supports the interaction plots. Despite the slight lack of parallelism in the lines for males and females, there is no evidence of an important interaction between housing and sex. There is little evidence of a main effect of housing, which is illustrated by the fact that the lines for both males and females are close to horizontal. The large distance between the male and female lines is indicative of a sex effect, which is highly significant in the ANOVA table. The types of behaviors observed occur more in female hooded rats than in male hooded rats.

Conclude. Female rats exhibit the three types of behavior counted more frequently than male rats. The size of the increase in these counts was approximately the same whether the rats were housed alone or in groups. The type of housing does not appear to effect the frequency of the social behaviors measured.